From its beginning at a long-abandoned and all-but-forgotten pioneer cemetery, this book sets off on a three-centuries-long odyssey through the Helen valley and the Chattahoochee headwaters. The region was the wild domain of panthers, Indians, traders, and soldiers for a hundred years until the Unicoi Road was begun in 1813. The next century belonged to the pioneers and gold miners whose story is told for the first time on these pages. The tale does not end there, though, for the writer follows the threads of the past to find them still woven into the evolving tapestry of modern life in the beautiful northeast Georgia mountains.

WHAT OTHERS SAY:

"Gedney weaves a captivating tale...a most refreshing approach which employs the best of fact-based history with the charm of good old-fashioned Southern storytelling. . . a beautifully done work -- I read it twice within a few days!"

> -- D. Michael Allison, President, National Allison (Allanson) Family Association

"A virtually unknown segment of history comes alive in this extremely interesting chronicle. From the England family who settled in the Helen valley in the early 1800s through the great gold mining era, this portion of Helen's history is re-lived through the very capable pen of Matt Gedney. The vivaciousness of his characters and the tales of the Gold Rush make this story as appealing as any book on the market today."

> —Helen Fincher, Executive Director, Helen-White County Convention & Visitors Bureau

"I saw this book born in the persistent archival research of this relentless writer. . . not only a well-researched but also a well-written glimpse into a pocket of 19th century Georgia civilization . . . some especially richly detailed history . . ."

> —Dale Couch, Historical Research Advisor, Georgia Department of Archives and History

LIVING ON THE UNICOI ROAD

HELEN'S PIONEER CENTURY
AND
TALES FROM THE GEORGIA GOLD RUSH

Matt Gedney

First Edition

Little Star Press, Marietta, GA

LIVING ON THE UNICOI ROAD

HELEN'S PIONEER CENTURY
AND
TALES FROM THE GEORGIA GOLD RUSH

By Matt Gedney

Published by: Little Star Press
175 Mt. Calvary Road
Marietta, GA 30064 U.S.A.

Design Concept by David Greear
Cover Picture Copyright 1996 by Ken Woodall and used by permission.

Printed in the United States of America

Publisher's Cataloging in Publication Data
Gedney, Matthias J.
Living On The Unicoi Road: Helen's Pioneer Century and Tales of the Georgia Gold Rush / by Matt Gedney—First Editio
   p. 192   23cm.
Includes bibliographical references.
1. Helen (Georgia)—History—19th Century
2. Gold Mines and Mining—Georgia

1996   LCC 96-75702
ISBN 0-9651196-9-6: $11.95 Softcover

CONTENTS

ACKNOWLEDGEMENTS

I had a lot of help. Thanks to: My parents, Leigh and Catherine, who in particular assisted in this effort in many ways. To wife Cindy, especially for reading everything and then facing demands to constantly re-read and see if those three words really needed moving. To brother Page for advice on geology. To Al Mills for being a kindred spirit and for insight into country living, particularly concerning oxen and mules. To David Robertson, who showed me around the Georgia Archives for the first time. To the helpful folks at the Archives, especially Dale Couch who found time to review and add considerably to this effort (even if he did insist on endnotes). To Martha Edge for information on her ancestral Englands and help with genealogy, her specialty.

To long-time local residents Comer Vandiver and Laura Abernathy Cannon, both of whom were around before Helen was and were very good in telling about it. To Laura's daughter, Emma Ruth Mize. Special thanks to Dr. Tom Lumsden who generously shared his years of efforts to preserve local history and is cited in too few places. To Walter Lumsden, also an accomplished collector of local lore, and to Isabel Couch for making introductions. To Vic Bristol, whom I never met but who in his unique and colorful way left a valuable window to the past. To Jim Vandiver, who has mined more Georgia gold than anyone else around. To Mitchell Vandiver for mining insight and years of hands-on experience with a shovel. To Mrs. Mary Davidson and her brother John Henderson, local descendants of the pioneer Conleys. To Ann and Bill Conley and Bill's sister Betty, all of the Texas branch of the family and keepers of the diaries of Sam Conley. To Jean Gilreath at the White County Library. To Sol Greear for stories on the Unicoi Road. To J.R. Dean descendant H. Dean Spratlin for materials on his family. To folk artist Ken Woodall for the cover picture, and to Linda Anderson for helping to bring it about.

To authors Jim Kloppel and Shelley Gill for advice on publishing. To the folks at Dukes Creek Woods for letting me revisit the Ben Fields and the old mines. To Melissa Craddock for my first chance to write. Particular thanks to David Greear for great work with photographs and fashioning the pieces into a complete manuscript. To the Mormons for copying all those official records. To Col. George Chicken, Dr. Matthew Stephenson, Sam Conley, Austin Dean, John Martin, Herbert Kimzey, and all the others along the way who for some reason left the writings and photographs which make it possible to obtain some sense of the way things used to be in a small stretch of valley on the headwaters of the Chattahoochee River.

PICTURE/EXHIBIT INDEX

* Courtesy Georgia Department of Archives and History.

THE MYSTERY OF THE OLD ENGLAND CEMETERY

When visiting Comer Vandiver a few years ago, I learned something interesting. Comer has been around the Helen valley longer than Helen itself. He remembers seeing the sawmill people talk with his father in 1911 as they scouted for a site to build the great mill which two years later gave rise to the town.

Today, Comer's house sits astride a narrow ridge which protrudes into the Chattahoochee river bottoms at the lower end of Helen, not far from the spot where he saw the sawmill men those many years ago. The circular driveway out front was once part of the old Unicoi Turnpike. Behind his house stands a huge beech tree. Measuring over 14 feet around at chest height, it's one of the largest in Georgia.

Scattered down the ridge below the ancient tree are about a dozen unmarked stones. At first glance, if you notice them at all, the stones look as if they've always been there, exposed by the forces of wind and water and left to weather atop the ridge.

However, as Comer pointed out, the stones are placed in an orderly fashion. And, if the great tree could talk, it could tell you a story, for the tree is witness to the fact that the rocks are really gravestones, simple markers which still proclaim the final resting place of some of the first settlers in the valley which became the city of Helen.

In recent years the old cemetery has been all but forgotten, remembered only by a very few long time residents like Comer. But its location was all that was known, since even the old-timers didn't know who was buried there. Although many local families have ancestors among the first settlers in the area, none seemed aware of any connection to the abandoned cemetery.

Comer did have a clue, though. Some years ago, a local historian named Vic Bristol had stopped by to tell him that "Englands, Bells, and Pitners" were buried there, in what was known as "The England Cemetery". Vic said he had a list, but he passed away in the late 1960s and it has never been found.

We were both curious to know something about these now mysterious people, but didn't imagine that much could be learned at such a late date. I was surprised a few months later, though, when a friend showed me around

1

Beside the ancient beech tree, Comer Vandiver points out headstones in the Old England Cemetery located behind his Helen home.

the Georgia Archives in Atlanta. There were copies of deeds, tax digests, military rolls, books kept by census takers, original surveys, wills — all kinds of records filled with intriguing clues about the past.

Starting with the names from the old cemetery, I looked for information on the short stretch of valley where the city of Helen was established in 1913. A lot of sorting and pasting was required, but it was exciting to see the early history of this small area gradually come into focus.

The three families identified by Vic Bristol were in fact among the first permanent settlers in the valley. A fourth family — the Conleys — emerged from the old records as well. It turns out that all of these families had been gone from the valley for many years before Helen was founded, and that is why the names of those in the cemetery were lost.

I eventually found descendants of all these pioneer families. Some lived within a few miles, but others were as far away as New Jersey and Texas. All had family histories which helped a lot. The Texas branch of the Conleys was especially helpful, for the diaries of their grandfather Sam Conley provided a detailed account of valley life in the years after the Civil War.

Early on, Martha Edge provided a lot of information on the pioneer Englands. Years before I learned of the England Cemetery, Martha had appeared at Comer's front door. An accomplished genealogist, Martha had relatives in neighboring Union County about thirty miles from Helen, not too far across the Blue Ridge. In researching the family tree, she had learned that one of her relations, a cousin in the England family from which she was descended, had married a Vandiver from Helen in the early 1900s.

But in talking to Comer, Martha learned something she never expected, for when he mentioned the "Old England Cemetery", she realized that her connection to the area went back much further. She already knew that some of her Englands had lived in the area, but hadn't known quite where. And now, entirely by accident, she had not only found the place where her ancestors had lived five generations before, but was in fact standing only a stone's throw from their final resting place. And in coming from the back side of the Blue Ridge, she had made a reverse migration of sorts, retracing a few small steps in the great American movement to the west.

At first, of course, the pioneer history of the valley is a story of the American frontier. It begins on Indian territory in 1813, when the Cherokees approved the construction of the Unicoi Turnpike, a wagon road through the heart of their Nation. On its way from the headwaters of the Savannah River to the "backwoods" mountaineer settlements in northeastern Tennessee, the

Turnpike ran through the Helen valley, which at the time was 30 miles inside the Indian boundary. Iron-rimmed wagon wheels rolled along this road, and in a few years the Indians were gone, replaced by white settlers.

In the isolating embrace of the Georgia mountains, the way of life established by these pioneers endured until a road of a different sort reached the headwaters of the Chattahoochee. In 1913, the trains of the Gainesville and Northwestern Railroad steamed in on rails of steel, providing a direct connection to what had become a very different outside world.

These events frame the pioneer century. After 1913, as an economic boom and the power of modern machinery imposed a new order on the small valley, the traces of the first settlers were swept away. Only the faintest of signs remain to suggest that the American frontier paused here for a while, delayed by the Cherokees in its rush to the west, and then left behind a hardy group of men and women who came with children and gave birth to more even as they struggled to build houses and claim their fields and pastures from a great wilderness. Today, a few creeks bear the names of early settlers and a few stretches of the Unicoi Road remain, but the Old England Cemetery is the only thing left which is known to have been fashioned by pioneer hands.

Although much of the early history centers on four families who came to own most of the property there — the Englands, Bells, Conleys and Pitners — many others passed through the Helen valley. Things were unsettled at first, for many of the first arrivals remained for only a few years before moving on once again to follow the departing frontier westward. Slaves were in the valley from the start, toiling in the fields and goldmines until the Civil War.

Plenty of white miners came too, digging along the creeks and hillsides and eventually using waters from great mining ditches run high on the mountainsides to flush the gold from its ancient hiding places. Helen's foremost miner turned out to be a migrating New Englander who found himself stuck on the Chattahoochee headwaters by a balky mule. And all the while, a parade of strangers passed along the Unicoi Road as valley residents tended their farms and engaged in enterprises such as the making of leather, the milling of corn, the weaving of cloth, and the hammering of hot iron into useful shapes.

All but two members of the first generation of permanent settlers found a final resting place in the England Cemetery or church cemeteries nearby. Most of their sons and daughters moved on, almost always in the westward direction, although a few are also found in local cemeteries. By 1890, the last of the pioneer settlers was gone. Most of those who took their place in the Helen valley were of local pioneer stock as well, comprising a sort of "second shift" which has many descendants in the valley even today.

These newcomers had no relatives in the England Cemetery. Whatever the old folks among them may have known about the old graveyard and the first settlers, it was apparently not much discussed with youngsters like little Comer Vandiver. Some of them do appear to have talked with Vic Bristol, though, for perhaps twenty or thirty-something years after his birth in 1882, he embarked on his self-appointed career as a chronicler of local events and came around to find someone who knew about the England Cemetery and the old days in the Helen valley.

And now, with thanks to Vic Bristol for his clue and Comer for passing it along, it's possible to say who these early people were. To a surprising degree, the curtain can be raised to look once again at times which have long since passed from living memory. Although sad events occurred at the England Cemetery, we also find pioneers arriving full of enterprise and optimism, dozens of children laughing and splashing in the icy waters of the river, and long-lived grandparents who must have been pleased with what they accomplished in this small stretch of valley on the headwaters of the Chattahoochee, where the old beech tree still watches over the traces of things they left behind.

"OUT OF THE HILLS OF UNION AND INTO THE LAKE THAT IS HALL. . ."

Before focusing on the Helen valley, it's helpful to take a quick look around, for its history cannot be told apart from its surroundings. At an elevation of about 1600 feet, Helen is 600 feet higher than Atlanta. The hills on either side are actually high points on two long ridges which extend for about nine miles to the crest of the Blue Ridge as they climb to heights of over 4000 feet. Over a century ago, gold miners ran ditches high along both of these ridges to divert the captured waters of neighboring streams into the Helen valley. The highest point overlooking the valley is Hamby Mountain, which stands along the ridge dividing the waters of the Chattahoochee from those of Dukes Creek to the south. Today, the WHEL radio tower stands on Hamby's summit.

Helen is on the headwaters of the Chattahoochee River, about 14 river miles below its high mountain source. Narrows at each end separate Helen from the Robertstown community above and the valley of Nacoochee below. Helen was not founded until a great sawmill was built there in 1913. Since the name "Helen" was not previously applied to anything local, references to the "Helen valley" prior to 1913 are made only to locate the modern reader.

The neighboring community of Robertstown lies just a mile above Helen. In fact, it's so close that it has sometimes been referred to as "North Helen", a designation which has never been acceptable to a true Robertstown resident. The place was named after a well-to-do Englishman who moved there in the late 1800s to make wine and mine gold. Robertstown was once a city, incorporated in 1913 the same as Helen, but the folks there soon decided municipal affairs weren't worth the trouble and the township was dissolved. The first white man to live there may have been named Smith, for when the original Georgia surveyor came by in 1820, he assigned Smith's name to the large creek which flows down through Unicoi State Park to join the Chattahoochee in Robertstown.

The storied "Vale of Nacoochee" begins just below Nora Mills. The name pre-dates not only the arrival of the white man, but apparently that of the Cherokees as well. Although a myth has arisen that Nacoochee means "evening star" in Cherokee, the word has no meaning in that language.[1] It probably originated with the Creek Indians, who appear to have been displaced by the Cherokees some years before the white man reached the area.

A gazebo-adorned Indian Mound sits at Nacoochee's upper end, one of four such mounds in this large valley. At its lower end, Sautee Creek enters from the adjoining valley of the same name, briefly crossing Nacoochee to merge with the waters of the Chattahoochee. In addition to the mounds and the Indian names, signs that the area was once a center of Indian culture include artifacts dating back many thousands of years and several latter-day Cherokee village sites.

Nacoochee was also a center of local pioneer life, as it was home to the first churches, school, stores, hotel, and post office in the area. The place was a little different from the start, for many of its first settlers trace back along a piedmont migration path rather than that of the Scotch-Irish/mountaineers more often associated with the Georgia mountains.

Nacoochee has always elicited a stream of effusive prose from enthusiastic tourists and verbose valley dwellers. One early resident was Dr. Matthew Stephenson, a learned man of considerable vision who nonetheless was known for exaggeration, often saying "millions" when those about him thought he should have said "thousands". In 1870, Dr. Stephenson reported Nacoochee to be a center of western civilization as yet untouched by the notions of Charles Darwin:

> The valley of Nacoochee. . . is considered by all foreigners to be one of the most charming and lovely valleys in the world. It is in a high state of cultivation, and improved by elegant residences, orchards, vineyards; and many of the choicest works of art in statuary and sculpture are being introduced from Italy to add to its beauty. Its inhabitants are natives of Massachusetts, Indiana, Virginia, Georgia, the Carolinas, and some from beyond the seas. They are pious, intelligent, and hospitable, and have yet to learn the cursed influence of European modern philosophy.[2]

Until the 1900s, "Nacoochee" was often used to describe a larger area encompassing neighboring valleys and habitations between the Blue Ridge and Yonah Mountain, the rocky prominence which overlooks it from the south. This area included the valleys of Dukes Creek and Sautee, and the Helen valley until it finally got a name of its own.

In a rather precise definition of importance, the Funk & Wagnall Encyclopedia calls Georgia poet Sidney Lanier "the outstanding southern poet of the last four decades of the 19th century"[3]. Although every Georgia school child is exposed to the great rhymer's works, most seem to have little recollection of the experience. A few can still recall the first words of one of his

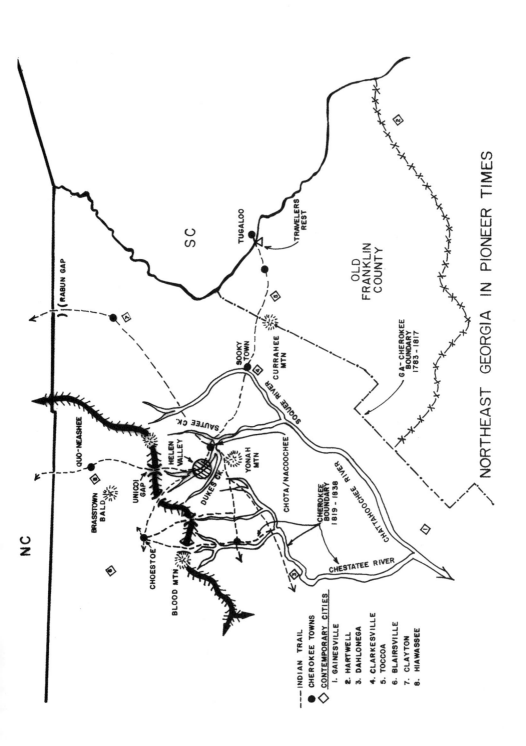

NORTHEAST GEORGIA IN PIONEER TIMES

most famous poems, "Song of the Chattahoochee", which are: "Out of the hills of Habersham and into the valleys of Hall..."

It's the sort of lyrical phrase for which Lanier was famous. However, there is a bit of poetic license involved. When the poem was written in 1883, the Chattahoochee River actually began in Union County and ran for miles through White County before reaching the hills of Habersham. It still does. And the river never really goes "through" the hills of Habersham, since the county only borders the river on one side.

"Habersham" just sounds good, though, and the poet seems to have favored the needs of his rhyme over the artifices of politicians and geographers. But then again, Sidney Lanier was not the first to relocate the Chattahoochee. Although the river itself hasn't moved, over the years it has been considered to begin in a number of different places, at least two of which were once located in Habersham County.

It's often said locally that "Chattahoochee" originated with the Cherokees, but in fact the name was first applied far downstream in the territory of the Creek Indians. According to unpublished notes made by U.S. Indian Agent Benjamin Hawkins in 1798, the name was taken from the oldest Creek town on the river, "Chatto ho chee", a village which was probably located in Georgia's Heard County not too far from the Alabama line.[4] As the name won acceptance among the whites, it steadily moved upstream, accompanied by a variety of spellings along the way.

However, for as long as the Cherokees remained on the Chattahoochee headwaters, the name only made it up to the edge of their territory. From the Cherokee perspective, the head of the river was known as "Chota River" or "Chota Creek", taken from their town of that name located in today's Nacoochee Valley.

Early maps had the Chattahoochee beginning about 25 river miles below its source high on the Blue Ridge, at the point where the Chota River merged with the Soquee River.[5] Later, an 1820 Georgia surveyor's map moved the start about eight miles further upstream to where the river was joined by Sautee Creek at the lower end of Nacoochee Valley. Above that point and on up through the Helen valley, the main channel was still identified as the Chota River.[6] Although both of these junctions were once located entirely in Habersham County, the county boundaries had been changed many years before Lanier wrote his famous poem.

One of the last references to "Chota River" appeared in Adiel Sherwood's 1837 Gazetteer of Georgia. Under the entry for "Chattahoochee

River", the Gazetteer said the river was "formed by two principal branches, the Chota and Sokee. . . which unite 8 miles below Clarkesville. The western branch now, however, is called the Chattahoochee, instead of Chota as it was by the Indians."[7]

With the departure of the Cherokees from the Nacoochee area, the memory of Chota quickly faded away. The map makers eventually traced the river to its source, following the larger stream at every junction to bestow the title upon a small spring which lies a short distance below the crest of the Blue Ridge. This spring was never in Habersham County. Instead, it was on Cherokee soil until the natives were finally forced out, and has been in Union County ever since. It's on the very edge of Union, though, running only a short distance before entering White County, which was carved out of Habersham County in 1857.

Although the early history might help the legitimacy of Sidney Lanier's rhyme, the years since have been less kind. The river runs pretty much the same through the hills of Union, White and Habersham, but the valleys of Hall are considerably different. For almost the entire length of Hall County, the Chattahoochee valley is covered by a large body of water known, appropriately enough, as "Lake Sidney Lanier".

On the Chattahoochee headwaters are two main tributaries which also have their sources high on the Blue Ridge. The eastern one is the aforementioned Soquee River. It takes its name from "Sookee Town", once located on its banks a few miles above the modern town of Clarkesville. This Cherokee village lay along an ancient trail which later became the Unicoi Turnpike, the main road from the headwaters of the Savannah River to the Tennessee settlements in the early 1800s. From Soquee Town, this trail ran through Nacoochee and the Helen valley on its way to Tennessee.

To the west is a larger tributary, the Chestatee River. According to Indian historian James Mooney, the name in Cherokee means "fire-light place", a reference to the "fire-hunting" method of killing deer in the river at night.[8] Letters written by pioneer settler William Jones say the name was applied to a Cherokee village on the river. Unlike the Chattahoochee and the Soquee, the Chestatee somehow never made it up to the Blue Ridge. Instead, maps have the river starting about six miles below the Ridge, at the junction of Frogtown and Dicks Creeks. The Chestatee was on the Cherokee boundary for about 20 years, from 1819 until the Indian Removal in 1838. It winds past Dahlonega, and once merged with the Chattahoochee below Gainesville. Today, the lower third of the Chestatee is covered by the waters of Lake Lanier.

UPPER CHATTAHOOCHEE. Near the intersection of GA 115/75, the gaze-bo adorned Indian mound stands at the upper end of Nacoochee Valley. The rooftops of alpine Helen appear in the middle of the picture. For nearly 20 years after the first settlers came, the lands across the Blue Ridge were the domain of the Cherokee Indians.

      The over-mountain counties of Union and Towns adjoin White County near the crest of the Blue Ridge. Although each contains a small slice of the Chattahoochee headwaters and Union even contains the river's source, most of their lands are on the back side of the Ridge and the Tennessee River drainage. Before Georgia embarked on a movement to create smaller counties in the mid-1800s, Towns was part of Union. The original Union County was created while its territory was still in the Cherokee Nation. With the encouragement of Georgia leaders, white settlers began streaming in several years before the Cherokee Removal. Some of these settlers were from the Helen valley, and Union County continued to be a surprisingly popular destination for them long after the Indians were gone.

Helen is located in the northeast corner of Georgia. As the crow flies, North and South Carolina are less than 25 miles away and the Tennessee state line is only about 40 miles distant. But since crows have a considerable advantage in the north Georgia mountains, the distances by automobile are nearly twice as far. Visitors from the South Carolina colony reached the Helen valley by at least the early 1700s. Georgians followed some years later, reaching the area no later than 1740.

Until the Civil War, northeast Georgia looked eastward towards Charleston, Augusta and Savannah for economic and political affairs. After the War, the focus turned to the west and the city of Atlanta, which at the time was a compact town 90 miles from the Helen valley. In recent years metropolitan Atlanta has grown northward at a rapid pace to occupy about half of the countryside which once separated the two. And, with Helen's rise as a tourist town, it's probably a safe guess that at least half of Atlanta comes to visit during the summer and fall seasons.

Visitors have always come to the mountains, though, even without the ease of modern transportation and the inducements of Helen's contemporary Alpine commerce. In fact, the first tourists weren't all that far behind the first settlers. There's just something special about the mountains, a sentiment which was well expressed by Dr. Stephenson in the flush prose of an earlier era:

> From the trending of the mountains the upheaved strata form thousands of waterfalls. . . in addition to which the charming landscapes, the magnificent scenery, and the sublime accompaniments of mountain grandeur, elevate the mind and fill the soul with the most exquisite and indescribable emotions, bringing man into close communion with his Creator and Preserver.[9]

DR. MATTHEW STEPHENSON. *Dr. Stephenson is usually associated with Dahlonega, but his Georgia gold career began in Nacoochee where he lived for years after purchasing interests in a number of local mines for $4000 in 1834. Although there are rival claims as to when and where north Georgia gold was first discovered, the colorful Dr. Stephenson always said the find was on Dukes Creek in 1828.*

TRAVELLING THE UNICOI ROAD

Except for a handful of whites who chose to live among the Indians, most of the Chattahoochee headwaters did not see a permanent European presence until the first settlers arrived in the 1820s. In Nacoochee Valley and the valley which would become Helen, however, white men had long been regular travellers along a noted trading path which bisected them. This access was formalized by treaty nearly a decade before the pioneers arrived, the story neatly summarized by the "UNICOI TURNPIKE" historical marker which stands near the Nacoochee Indian Mound to proclaim:

> This is the old Unicoi Turnpike, the first vehicular road to link eastern Tennessee, western North Carolina, and north Georgia with the head of navigation on the Savannah River system. Beginning on the Tugalo River to the east of Toccoa, the road led through Unicoi Gap, via Murphy, North Carolina to Nine Mile Creek in Maryville, Tennessee. Permission to open the way as a toll road was given by the Cherokees in 1813 to a company of Indians and white men. Georgia and Tennessee granted charters to the company.

It took the Unicoi Turnpike Company a while to complete the road, but by 1819 it had become "the great highway from the coast to the Tennessee settlements".[1] According to one early account, a four horse wagon could carry 2500 lbs. without difficulty and make the trip from eastern Tennessee to the Savannah River trading town of Augusta in 10 days less time than before.[2] "Turnpike" derives from the type of gate which marked the start of such roads, typically a long pole or "pike" which was turned aside when the toll was paid.

Although it may seem an early effort, the Unicoi Road was not Georgia's first transportation venture into the Indian region. By 1740, the young Georgia colony had opened an improved trading path for horsemen out of Augusta.[3] It became much used, a considerable improvement over the previous state of affairs where most goods were carried down narrow trails on the backs of Indians. The trading path ran up the west side of the Savannah River to Cherokee towns on its headwaters, intersecting with the trail that became the Unicoi Road in the vicinity of Toccoa. This improved path was necessary if Georgia traders were to compete with rival merchants in the South Carolina colony, who had long profited from several improved routes to the west.

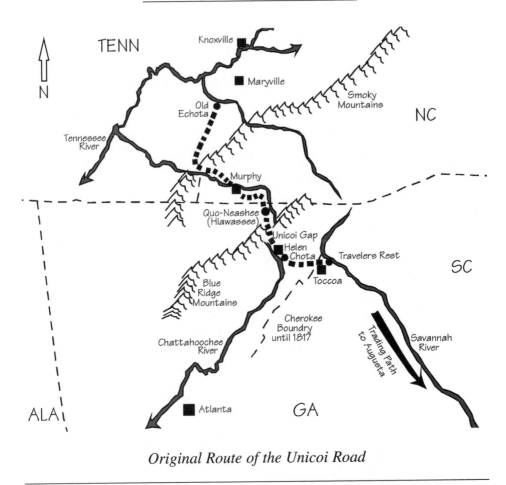

Original Route of the Unicoi Road

"Unicoi" derives from the Cherokee word for the color white, so that to them the turnpike was the "white road".[4] Today it's not known in what sense the name was applied, but there are several long-standing theories. The most common is that "white road" refers to its use by the white man. Another has the name arising from the winter snows or frosts which sometimes whitened the high areas along its length, an explanation which may relate to the fact that "Unicoi" is also applied to a high section of the Great Smoky Mountains where the old road crossed the Tennessee-North Carolina border. A less often heard story attributes the name to the several "white" or peace towns which once lay along the route, among them the abandoned town of Chota in Nacoochee Valley and the Cherokee capital of Echota at the northern end.

Locally, the Unicoi Road ran the length of both the Nacoochee and Helen valleys. After the grading and re-aligning associated with modern high-ways, only a few sections remain visible near Helen. The old roadbed can most easily be seen at the south end of town, where it is dug into the bank beside the Highway 75 curves just above Nora Mills. Helen's Main Street follows the old route, except that early travelers had to ford the river where the Main Street bridge is now.

At the upper end of Helen, Highway 75/Main Street and the old route part ways. While the modern highway has been blasted through rocky bluffs to stay on the east side of the Chattahoochee as it leaves town, the old road of necessity took a gentler route, crossing the river at a second ford in front of the present Orbit Manufacturing Company to exit the Helen valley along the route of today's Ridge Road. Laura Cannon lived for many years just above this second ford. Looking downriver towards town from her perspective at the upper end of the Helen valley, this is how she remembered the Unicoi Road in the early 1900s:

> [the road] went right along the edge of the river — a little narrow road — and you went just right along the river, and down here right in front of the house was where you forded the river, and you went down and you came out at that eating place [the Hoffbrau House]. . . if you was in a wagon, buggy or anything, you forded the river down here in front of my house and kindly went down the river and out at that [eating] place. . . . The main street, that was just a little dirt road, that road was there. They was a bridge after so long a time [the Main Street Bridge in the middle of town], but they was a long time they forded the river again. They had a swinging footlog that the people could walk across, but the wagons had to ford the river, again.[5]

Ten miles north of the Helen valley, the route crossed the Blue Ridge at Unicoi Gap, the lowest spot for miles in either direction. Even so, at 3000 feet in elevation it presented a formidable obstacle for road builders and travelers alike, requiring a climb of 1200 feet in less than three miles in the final approach from the Helen side. Of course, those going in the opposite direction faced a steep decline. Since wagons lacked good brakes, a heavy load could make for dangerous going. Wagoners sometimes remedied this problem by tying on to a good-sized log and dragging it behind to slow their descent.

Unicoi Gap is today traversed by both Highway 75 and the Appalachian Trail. Although the gap itself has been bulldozed several times for road improvements, the old Unicoi Road can readily be found headed down the mountain on either side. While the south or Helen side of Unicoi Gap drains into the Chattahoochee River, waters falling on the north side of the gap start a long journey to the Mississippi and Gulf of Mexico via the Hiwassee, Tennessee and Ohio Rivers.

OLD AND NEW ON THE UNICOI ROAD. This 1930s picture was taken at the upper ford shortly after the modern highway was blasted into the river bluffs on its way up to Robertstown and Unicoi Gap. Fill from the new road blocked the old Unicoi route, which entered the river just above former Orbit Manufacturing and came out again about where the Hofbrau Haus is today. Made locally, the little wagon had been in the Helen valley for some years. The "tall, raw-boned mountaineer" standing behind is Henry Newton Abernathy. Anna and Leigh Gedney are in the wagon.

The Unicoi Road was preceded by an ancient trail, a major southeastern trading path on the extensive network of footpaths which criss-crossed the American wilderness for thousands of years before the white man came. In January of 1716, Colonel George Chicken came up from the South Carolina colony and traveled this trail between the town of Chota in Nacoochee Valley and the Cherokee settlement of Quo-neashee (located near present day Hiawassee[6]) across the Blue Ridge. Chicken found it a rough trip, for he wrote:

> Sunday Ye 22} this day at 8 a clocke we seatt out from Chottee [Chota] to go to Quo-neashee we marcht about 20 miles ye way verry mountannas and stoney being forct to light and walke more then ride then we come to ye tope of ye mounton [Unicoi Gap] and ther we see the hade of a River that Rones in to Chattahouchey River about a mile one ye other side of ye mounton ther begon ye hade of a nother River [Hiwassee River] that rones into masashipey ouer march this day was 40 milles wee come to Quoneashee 1/2 hower after five a clocke where River that we see ye had of was verry brode[7]

Probably due to the difficulty of the route, Chicken judged the distance from Chota to Unicoi Gap to be nearly twice as far as it actually was. However, his knowledge of geography was outstanding, for other early visitors typically confused the Chattahoochee headwaters with those of many other Georgia waterways. Such mistakes were easy to make in an uncharted wilderness, particularly in northeast Georgia where narrow valleys give rise to a number of rivers heading in all directions.

Fifty years later, the same trail was travelled in the opposite direction by South Carolina militiamen during the Revolutionary War. The South Carolina Army first pushed through Rabun Gap to attack Cherokee settlements across the Blue Ridge around present day Franklin, NC. As the rampage continued, towns on the headwaters of the Hiwassee River were also destroyed, including the town of Quo-neashee which Col. Chicken had visited 50 years before. From there, a detachment was sent southward to re-cross the Blue Ridge at Unicoi Gap and attack Chota and neighboring villages on the Chattahoochee headwaters.

Leaving the ruins of Quo-neashee, the soldiers spent the night at the foot of the mountain on the north side of the Blue Ridge. They were apprehensive as they approached the narrows around Unicoi Gap the next day:

> Tuesday the 1st day of October [1776], we prepared to march. . . . Our orders, when attacked, was for the two wings, that is to say, right and left, surround the enemy, and our division fight our way into the front. These orders we willingly consented to, and were ready to obey when occasion served. We marched with all care possible, sending out flanking guards on the mountains, thinking we should have an engagement at the head of Highwassah waters [Unicoi Gap], on account of the narrows thereat, and the mountains on both sides, which on account of the pilot [Indian guide], was as convenient for them as Black Hole [site of a previous ambush]: But with great courage and resolution, resolving to have satisfaction, or die in the attempt, for the great slavery and hardships they put us unto, and more particularly, the loss of so many gallant men, we marched up the mountain. . .[8]

The steep, laurel covered mountains rising on either side did present their Cherokee foe with a considerable opportunity, but there was no ambush at the gap. Unmolested, the Carolinians continued on down the mountain to destroy Chota, complaining about the sixteen stream-crossings they had to make along the way.

Conflicts with the Cherokees continued until a permanent general peace was finally established in the mid 1790s, allowing for a degree of cooperation between the two cultures. In this spirit, Georgia expressed formal interest in the Unicoi project by at least 1812, when the General Assembly requested federal approval for the road. The next year, in March of 1813, a treaty was negotiated at the U.S. Cherokee Agency in Tennessee for the construction of the Unicoi Turnpike through the Cherokee Nation.[9] It was to be run by the Unicoi Turnpike Company, which would pay the Cherokees a fee of $160 per year for 20 years, after which time the road would revert to the Indians unless other arrangements were made.

The road was subsequently chartered by both Georgia and Tennessee. When Georgia passed its authorizing legislation in 1816, Unicoi became the first turnpike in Georgia to be so approved. Tolls were established as follows:

> . . . the company shall be entitled to receive the following tolls and rates at said turnpike, for the passage of any person or thing, that is to say: For every man and horse, twelve and one-half cents; for every led horse not in a drove, six and one-fourth cents; for every loose horse in a drove, four cents; for every foot man, six and one-fourth cents; for every waggon and team, one dollar; for every coach or chariot, or other four wheel carriage, chaise, chair or other carriage of pleasure, one dollar and twenty-five cents; for every two wheel carriage, chaise, chair or other carriage of pleasure, seventy five cents; for every cart and team, fifty cents; for each head of cattle, two cents; for each head of sheep, goats, or lambs, one cent; for each head of hogs, one cent.[10]

The road had to be at least 20 feet wide except where bridges or digging were required. Then, a minimum width of 12 feet was allowed, with the center of the road to be cleared of rocks and stumps, "if the same will permit". Since the road was built with only hand tools and animal power, the stumps and particularly the rocks did have considerable say in the routing of the road.

The road could not exactly follow the old footpath. While foot travelers were free to thread their way through the rugged terrain, a wagon road had to be wide and flat. With their limited means, the road builders had to skirt obstacles the trail had traversed. The effect of this showed up between Nacoochee Valley and Unicoi Gap. Where the Carolina soldiers in 1776 had noted 16 stream crossings on this section of the path, the road builders were forced into 28 along the same stretch, and in places simply went down the middle of the creek when there was no good alternative.[11]

Although the company sent a note to the federal Indian Agent in March of 1814 informing him that work on the road was about to start, construction through the rugged Appalachians took longer than expected. The Georgia legislature initially required completion in 1817, but the date was later moved to November of the following year. The story was similar at the Tennessee end, where the legislature also granted extensions. The road was finally in full operation by 1819, when it was advertised as a safe route "with as much convenience as any other road through the Cherokee Country"[12] (there were few competitors for this claim).

Early on, the turnpike seems to have experienced financial difficulties. In 1821, the Georgia legislature authorized a loan of $3000 to the "Unacoi" Turnpike Company with a term of five years. It was not repaid on time, for in 1826 the Governor directed the Comptroller General to go out and collect the money. The Cherokees also had troubles with the company, complaining in 1825 to the U.S. Agent about the failure of the Company to make the promised annual payments and making a similar protest two years later when the promised sums were still unpaid.

In spite of the financial problems, operation of the road continued. Roadhouses were located every twenty miles along the turnpike, this being the average distance a horse and wagon could make in a day. One of three roadhouses operated by General James Wiley was located at the lower end of Nacoochee Valley. Offering hot fires, cooked meals, and warm cups of coffee to savor along with the company of fellow sojourners and the usually sociable innkeepers, the roadhouses were havens for the weary. The beds were dry, even though it might be necessary to share one with a stranger. Outside, food was available for the stock; sometimes a drover might leave part of his herd as payment for the hospitality he received.[13]

All of the vehicles and animals contemplated in the Georgia Act of 1816 passed along the road in considerable numbers, but there were other travelers as well. Among the most comical were large flocks of turkeys, cantankerous critters for which the Georgia legislature had failed to set a toll. Although they could be driven down the road, doing so required patience, for a flock of gobblers could make only about 8 miles a day. In the evening, it was the birds themselves rather than the drover who decided when it was time to roost. And once they did, nothing could dislodge them until the sun reappeared the next morning.[14]

Gold miners, peddlers, photographers, entertainers, adventurous writers and hardy tourists; all of these passed along the Unicoi Road. The drovers with their flocks and the purveyors of furs and leather headed south, bound for markets at Augusta and beyond. Profits from these exports brought manufac-

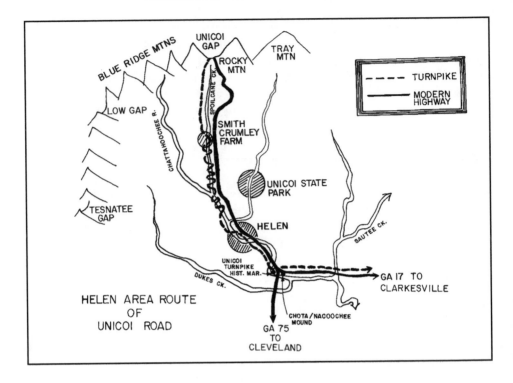

HELEN AREA ROUTE
OF
UNICOI ROAD

tured goods and necessary supplies northward. Settlers came from both ends of the Turnpike as the Cherokee areas along its length were opened for settlement. Even gypsies traveled the Unicoi Road. With colorful dress and ornate wagons in old world style, they did business in traditional crafts like tinsmithing. They were also good horse traders, perhaps even a little too good, as the locals felt it necessary to keep an eye on their own horses when the gypsies were around.[15]

Although the Unicoi Turnpike was over 150 miles long in its early years, it began to shrink soon after it was completed. Treaties with the Cherokees in 1817 and 1819 put the first 40 miles of the road on Georgia territory, moving the Indian boundary from the vicinity of Toccoa and the Tugalo River westward to the crest of the Blue Ridge. The road then reached the Cherokee Nation at Unicoi Gap, which remained the entry point until the Removal in 1838. As the Indian boundary migrated west, the turnpike marking the start of the road dutifully followed at some distance behind, and for a time was located in Nacoochee Valley.

In the late 1830s, when the charter of the original Unicoi Turnpike Company lapsed and the Cherokees were removed, the State took over maintenance of the remaining Georgia portions of the route, about 40 miles in all. Road commissioners were appointed and funds appropriated for the improvement of the road from Nacoochee Valley to the North Carolina line. One of the commissioners during this period was Martin England, a pioneer settler in the Helen valley who had since followed the Unicoi Road across the Blue Ridge to settle on the other side of the mountain. Another was Adam Pitner who lived beside the road in the Helen valley.

A few years later, the Georgia portion of the Unicoi Turnpike again became a commercial enterprise. Except for a short break during the late 1800s, operations continued in some fashion for another 80 years. As the area developed, county governments took over most of the route, so that after the Civil War the remaining section operated as a turnpike was less than 10 miles long, extending only to the foot of the Blue Ridge on either side of Unicoi Gap.

When commercial operation of the Unicoi Road lapsed for a time before 1900, a new and competing turnpike was constructed across the flank of Tray Mountain a few miles to the east. However, when the new turnpike itself fell into disuse, the Unicoi Turnpike was resurrected one last time, continuing in operation into the 1920s. The last gate was located beside Spoilcane Creek, on the Crumley place at the foot of the mountain (the site is 6 miles above Helen on GA 75). Spoilcane is the Chattahoochee tributary which descends from Unicoi Gap. Its interesting name probably derives from events associated with the old trading route.

Dr. John Goff suggests that Spoilcane Creek was named for a minor ecological disaster which occurred there.[16] He starts by noting that "reed" or switch cane was once plentiful along all the waterways in Georgia. An ancient trading path had long run along the entire length of the creek which became Spoilcane, but even before it became the Unicoi Turnpike, not only had the route become more heavily used, but more importantly, horsemen and packhorse trains became frequent travelers along the trail.

Early travelers through the Georgia wilderness found very little grass or corn, but they could usually count on finding wild cane along the trail as food for their animals. However, as Dr. Goff put it, "The traffic up the narrow defiles of Spoilcane, however, was so heavy, it is reasonable to conclude the supply of cane along it was very early "spoiled", i.e. overgrazed or used up."

And, since the lower half of Spoilcane is relatively flat, it was a natural stopping place for those making the steep traverse through Unicoi gap, a factor which also would have contributed to overgrazing. "More than likely, the designation Spoil'd Cane was a well understood, precautionary expression which warned travelers to expect little forage for their animals if they proposed to stop along the stream for the night on the long tedious climb up to the gap." Even today, there is little cane growing along Spoilcane Creek.

Sidney Smith Crumley was the last full time gate-keeper on the Unicoi Road, manning the turnpike until about 1925.[17] The Crumley place was about halfway up Spoilcane Creek, occupying the last good bottom areas at the foot of the mountain. Set amid fields first farmed a century before, Smith's two story log house sat on the upper side of the road, overlooking the gate. Across the road, a barn built of rough wooden slabs adjoined a fenced yard where stock could be kept for the night. Smith operated a small grist mill for which he'd also charge a toll.

Comer Vandiver is Smith Crumley's grandson, and remembers the Turnpike in its final years. At that time, most of the people using the road came from the vicinity of Brasstown Bald and the town of Hiawassee, areas not too far on the other side of the Blue Ridge. They were headed for markets about twenty five miles from the Crumley place, in the towns of Clarkesville and Cornelia where there were rail connections.

As best Comer can remember, it cost 25 or 50 cents for a wagon. Traffic was the heaviest in the fall of the year. Hogs were killed then since the weather was cool and they wouldn't spoil as fast. Comer recalls seeing wagons loaded with "8 or 10 old dead hogs on them", all appropriately "dressed" of course. Fall was also harvest time, when the wagons carried loads of apples and chestnuts. One man would come from over the mountain, selling chestnuts along the way, a cupful for a nickel. As Comer tells it, "He had a big thumb, he'd stick it down in the cup, so you bought his thumb too when you bought a cup of chestnuts."

While wagons were still the main thing when Comer was young, new vehicles had joined them on the Unicoi Road. A number of families lived along the lower reaches of Spoilcane, where there was a school and a church, and one enterprising resident wrote a weekly column for the county paper. It was originally entitled "Spoilt Cane Dots" but later changed to "Spoilt Cane Special" because the columnist decided that "Special is better suited to the place".[18] Among the tidbits for the week of June 22, 1917 was the following: "The first auto on Spoilt Cane for 1917 made its trip Monday". Although automobiles had been cruising city streets for nearly two decades, they could still make it up to Spoilcane only when weather permitted.

For a brief period after Smith Crumley's watch, a woman came up from the nearby village of Robertstown to oversee the gate part-time, but the route was again abandoned as a toll road by 1928. By this time, though, the days of private turnpikes were just about over as the State had assumed responsibility for major roads. But since it took the Georgia highway department a while to reach Unicoi Gap, the old route was maintained by the newly arrived Forest Service, which wanted access to the forests in case of fire.

A Forest Service work crew under the direction of Ed Hollingsworth employed bulldozers to repair the road in 1928-9. During this period, Verge Adams ran the mail between Helen and Hiawassee, a job which required regular trips through Unicoi Gap. Sol Greear of Helen accompanied Verge on one of these trips.[19] It was still rough going in Verge's T-Model Ford, for it took them several hours to cross the mountain. Sol described the scene at the gap, where they stopped to get chestnuts. Icy weather had cracked the spiny burrs, making the chestnuts easy to get. They had to beat the wild turkeys, though, for there were flocks of them living in the gap and the chestnuts would belong to the gobblers if they got to them first.

In 1939, using convict labor, State highway crews finally made their way up Spoilcane Creek and on across the mountain to open the first generation "free road" through Unicoi Gap, although it would be some years before it was paved. With the power of modern explosives and machinery, the highway engineers were free to choose a different route, particularly along the upper reaches of Spoilcane, where they climbed along the east side of the creek and left the old roadbed intact on the other side as they approached Unicoi Gap. In the years since, the new road has been paved, and then straightened and widened and paved again. The prolific chestnuts were killed by an imported blight and most of the turkeys are gone from the gap. On the lower reaches of Spoilcane, kudzu now grows where the original Georgia cane is said to have been.

In its heyday, the Unicoi Turnpike was as important as any modern freeway. While the turnpike itself seems never to have been a big commercial success, the road was vital to the early Georgia economy, for the State did not hesitate to lend financial support or step in and take over when the original turnpike company lapsed. Without the road, Georgia merchants in Augusta and Savannah would have lost business to their competitors in the Carolinas. In over a century of use, the old Unicoi Road saw everything from the soft footfalls of red men and the hard faces of iron-rimmed wagon wheels to the narrow rubber tires of early automobiles and the clanking treads of Forest Service bulldozers.

Although many mountaineers lived far back in the "splendid isolation" of the Appalachians, this was never entirely the case for those who lived along the Unicoi Road. Like the Gold Rush, it brought in outside influences and was one of those things which made folks in the Helen and Nacoochee valleys a little different from their neighbors. Along with the wagons and the droves came a parade of strangers and a steady supply of the latest news. And, since it preceded the first settlers by some years, it was the route which steered many of them to the lovely valleys on the headwaters of the Chattahoochee.

SPOILCANE DEFILE. Following the route of an ancient footpath, the Unicoi Road crossed the Blue Ridge at Unicoi Gap, the lowest point for miles in either direction. It ran along Spoilcane Creek, which makes a steep descent from the gap to join the main channel of the Chattahoochee about three miles above the Helen valley. The clearings in the foreground are the old Smith Crumley Place, site of the last gate. The modern highway can be seen snaking up the ridge overlooking the defile from the right.

OF "BEING FORCT TO LIGHT AND WALKE MORE THEN RIDE. . ."

July, 1990s. At 11 AM on a sunny weekend day in Helen, it's already nearly 90 degrees, hotter than it should be, and the tourist crowd is building. It takes only about twenty easy minutes to drive the 10 miles from Helen up to Unicoi Gap. There, the thermometer says it is 78 degrees, not a tremendous difference but certainly a pleasant one. It's humid, though, as dampening mists still hang in the air above the creeks which arise on either side of the gap. It rains more in Helen than it does down in Atlanta, and rains even more along the crest of the Blue Ridge, 20 or 30 inches a year more than down in the piedmont.

Down by the misty creeks, it's easy to find the old Unicoi Road. Or it might be better to say "roads", since in many places there are several old roadbeds running side by side, the newer lanes built when it became easier to just move the whole thing over a few feet than to repair the ruts and washes in the previous path.

There is an underlying order in the natural environment, but to the eye, things can look kind of ragged. The north side of the gap shows signs of a recent windstorm. Trees are down, a profusion of undergrowth thriving in the newly admitted light. Insects flitter about and boulders are strewn where ancient upheavals, floods, and the effects of ice have left them. In contrast, the old Unicoi Road has a character which only man can impart. As flat and straight as its builders could make it, the road marches over the mountain and through the gap, full of determination and purpose, for it once had places to go and a great wilderness to cross.

Back in the gap, about a dozen cars are parked, their occupants dispersed along the Appalachian Trail. On the edge of the parking lot, a sweaty and tired-looking young man sits in the shade. Beside him stands his bike, a fancy one with many gears, saddlebags, and a big water bottle. At the moment, the water bottle is the most important part. He is riding south to Helen and wants to know if this is the top of the mountain.

Although it must rise to the same height, Highway 75 is not nearly so precipitous as the Unicoi Road. The old route had to keep close to the creek, which got real steep as it approached the gap. The modern road, though, cuts into the side of the mountain near its base, starting the climb early to ease the ascent. Even so, and even with all of his gears, the young man says the climb was difficult, hard enough to force him off his clever machine and into walking a good part of the way. This, of course, is by no means a new complaint. In fact, its a pretty good echo of a similar one made nearly three hundred years before by Col. George Chicken, when he ruminated into his journal about "being forct to light and walke more then ride. . ."

"THE DWELLING PLACE OF THE PHANTOM
THINGS THAT WERE. . ."

When the first settlers arrived on the headwaters of the Chattahoochee in the early 1820s, over two hundred years had passed since the English established the Virginia settlement of Jamestown. Although there were white settlements further to the west elsewhere on the American conti-nent, the Cherokee Indians had held on in the Southern Appalachians and most of northeast Georgia had been theirs until the treaties of 1817 and 1819. When the area was finally opened for settlement, white occupation didn't take long, for there were large groups of frontiersmen nearby who had been steadily making their way towards the Georgia mountains for well over a century.

The great migration to America was, as it continues to be, a blend of many smaller migrations. Two groups of immigrants who began arriving in the 1600s were of particular interest in the settlement of much of Georgia, even though they first put ashore many hundreds of miles north of Georgia soil.[1] Encompassed by the colonies of Virginia and Maryland, the Chesapeake Bay region was settled largely by Englishmen. In this area a plantation econ-omy developed, one which shaped the "Gone With The Wind" vision of the South. Using slave labor, large farms used produced exports which were trad-ed for manufactured goods made elsewhere.

Further north, in Pennsylvania and adjacent colonies, the Delaware River valley was settled by a mixed group, English at first and then Germans, but predominately Scotch-Irish after 1730. Secondary emigrants from this group eventually came to occupy much of the southern Appalachians, inhab-iting a rugged region which required an economy much different from that of the flatlanders. Small highland farms were typically worked by their owners, who had little to export and therefore had to have a high degree of self suffi-ciency.

By the early 1700s, both the Chesapeake Bay and Delaware River groups had grown to the point where they gave rise to their own migrations. Taking parallel paths, both groups expanded southward to reach Georgia in significant numbers well before 1800. As the Chesapeake Bay people moved down through Virginia and the Carolinas to Georgia, they skirted inland from the already settled tidewater areas along the Atlantic coast to occupy the upper coastal plains and much of the piedmont region.

The Delaware River group grew westward until it bumped into the Appalachian Mountains in the 1730s. Turning southward, their migration pushed down along the mountains to occupy the valleys and foothills of the Appalachians. Although heavily Scotch-Irish at first, this group changed along the way as members of various ethnic groups joined the movement. Some new participants were of later generations born in America, but there were newly arrived immigrants as well, many of whom disembarked at more southern points down into the Carolinas and came from there to the Southern highlands. Many old families in the Helen area still bear names derived from these highlanders, representing English, German, Scotch-Irish, French, and even Dutch ancestry.[2]

By the time of the American Revolution this mixed group occupied extensive areas in the western Carolinas and the adjacent portion of Georgia, and a branch had crossed the Appalachians into northeast Tennessee to push southward to the areas around present day Knoxville. These mountaineers were described in epic fashion by Theodore Roosevelt in his Winning of the West:

> In this land of hills, they took root and flourished, stretching
> in a broad belt from north to south, a shield of sinewy men
> thrust in between the people of the seaboard and the red war-
> riors of the wilderness. . . a single generation, passed under
> the hard conditions of life in the wilderness, was enough to
> weld together into one people the representatives of these
> numerous and widely different races. . .[3]

Although these are the basic patterns, neither mountaineers nor piedmont people alone explain the settlement of the Helen/Nacoochee area, for the streams of migrants were mixed when they reached the headwaters of the Chattahoochee. Local families can be traced back along both of these routes and the early settlers showed both patterns of living. As they did elsewhere in Georgia, those of the Chesapeake Bay/piedmont culture tended to occupy the best lands. Coming with money and slaves, they were soon managing large farms in the fertile bottom areas along the Chattahoochee. In the smaller valleys on the tributaries and back in the mountain coves, family farms were established in the Scotch-Irish/mountaineer tradition. To a noticeable extent, descendants of each of these groups can still be found in the areas first occupied by their ancestors.

And even in the increasingly homogenized modern world, the echoes of these sweeping American movements can still be heard. That "southern drawl" voiced in much of middle and lower Georgia owes a lot to the Chesapeake Bay people, while the residual Appalachian "twang" in the north

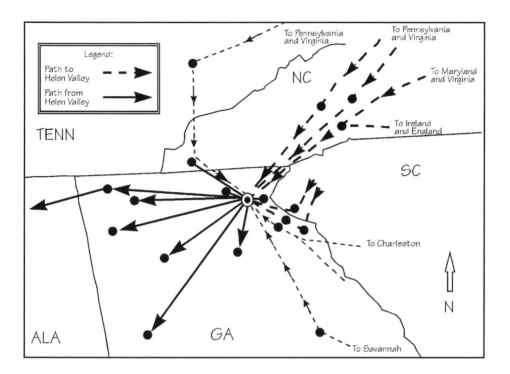

MIGRATION PATHS. The Helen area was opened for settlement in 1820. Most settlers came from two areas: the first came from the nearby Georgia counties of Habersham and Franklin; pioneers from the North Carolina foothills started arriving soon after. Some early arrivals came from other areas such as middle Georgia, South Carolina, and down the Unicoi Road from the backwoods settlements in Tennessee. Migrants came from both the piedmont and mountaineer migrations which had originated in Pennsylvania, Virginia and Maryland and which were joined by immigrants arriving at more southern ports as they pushed south and west. The frontier was delayed in the Helen valley for over a decade by the Cherokees. When it opened again, many pioneers picked up and followed it westward, as did many children of the permanent settlers who remained in the area.

is an enduring legacy of the Delaware River Valley group. In Georgia, it seems that the "twang" has been more pronounced on the back side of the Blue Ridge. And, on the front side of the ridge, while traditional Appalachian speech patterns have always been strong on the upper Chattahoochee, they were leavened by piedmont influences from the start.

For having sided with the British, the Cherokees lost considerable territory at the close of the Revolution. And although they were forced into further concessions in the decades that followed, the Cherokees nonetheless hung on to block settlement across much of the southern highlands for another half century. In Georgia, the Cherokee boundary after the war was fixed about 30 miles southeast of the Helen valley, where it stayed for nearly 40 years. The line was officially described as the "Western Frontier" and the new county of Franklin established on the border. Until the early 1800s, this was a dangerous area, subject to raids perpetrated mostly by Creek Indians but with some depredations by the Cherokees as well. In response, 14 log forts were built along this line, small fortifications into which local families could retreat if time permitted.[4] One of these structures is still standing.

The Indian problems had been resolved by the time the Cherokee treaties of 1817 and 1819 finally opened most of the Chattahoochee headwaters to white settlers, so no forts were required in the Helen/Nacoochee area. For two decades after that, though, the Cherokees were never far from the Helen valley. Until the Removal and Trail of Tears in 1838, they occupied areas less than twenty miles away across the Chestatee and just over the Blue Ridge, making occasional visits to their former territory. About a half a day's ride to the west, Fort Scudder on the upper Chestatee is said to have been one of the stockades used as a collecting point for the Trail of Tears.[5]

Among the early settlers were men who had served in the last generation of Georgia Indian fighters. A few of the older ones had fought against the natives in the Revolution and the last battles with the Cherokees shortly afterwards. Others had served in the later wars against the Creeks, when the Cherokees fought on the American side. Although memories of conflict were fresh and the sight of visiting Indians caused a stir on the upper Chattahoochee, hostilities were never a real concern since the Cherokees had finally chosen the path of peace several decades before the pioneers arrived. Even when sad events connected with the Removal took place only a few miles away, there were no Indian troubles for the newcomers.

When the first settlers came to the Chattahoochee headwaters during the early 1820s, the signs of the former occupants were clear. The Indians left a well established network of trails. The luckiest pioneers reclaimed the abandoned fields of the natives and perhaps even occupied a few of their houses.[6] As local lore has it, they may have found a few Cherokee stragglers as well, but it appears that most if not all of the natives were gone — and traditional Cherokee culture long gone — by the time the settlers arrived.

As it was all across America, the changing of cultures was a poignant moment. If any of the pioneers stopped to think about it, they may have seen things as Georgia historian Charles Jones put it some years later

The years roll on, and an increasing population, overleaping stream and mountain barrier, fills the hills and valleys of a dis-

tant interior. Before its inexorable advance the red race retires.
. . and the locality becomes the home of departed memories,
the abode of traditions, and the dwelling place of the phantom
things that were. The same bold river with restless tide has-
tening onward to mingle its waters with the billows of the
[sea]. . . , the same overarching skies, the same potent sun, kindred forests and voices of nature, but all else how changed![7]

And the settlers did come to a land of kindred forests and voices of
nature, following the Unicoi Road into the heart of a great mountain wilderness. While buffalo had once lived within at least 30 or 40 miles of the Helen
valley and elk almost certainly roamed the Georgia mountains as they are
known to have done in Alabama and the Carolinas, these animals were gone
many years before the settlers arrived. Wolves and panthers were not, however, and their fearsome cries echoed through the virgin woods.

The average tree was larger and there was less undergrowth in the pri-
mal forest. One of every four trees was a chestnut, a prolific food source
which supported an abundance of wildlife and some of the settlers' animals as
well. The only trout in the streams were native brook trout, but they were
plentiful. Extensive cane-breaks were found along the creeks, many of which
ran clearer and colder than they do today.

Armed with tools of iron, the newcomers started cutting on the edges
of this great wilderness. The panthers endured for a while, but the wolves
soon joined the elk in local extinction. Methodist and Baptist preachers held
forth where native shamans once had been the keepers of sacred rituals. With
the arrival of the pioneers, things were certainly different in the dwelling place
of the phantom things that were.

THE PIONEER FAMILIES OF THE HELEN VALLEY

With the Cherokees having retreated some distance to the west, eager newcomers were soon on their way to the Helen/Nacoochee area, the poorest traveling via foot power and packhorse while the richer ones came with their wagons and slaves. Most came from nearby Georgia counties like Franklin which had been settled for over four decades and from areas in upper North Carolina around the present day cities of Asheville, Hendersonville and Morganton. Some came from the South Carolina highlands and a few came down from the Tennessee branch of the evolving Scotch-Irish/mountaineer migration, which had direct access to northeast Georgia via the Unicoi Road.

Looking backwards along the migration paths of the Helen valley pioneers, their movements show a generational pattern, where a family typically moved once or twice to settle in a virgin area, and there the parents remained to watch many of their sons and daughters disappear over the ridge, bound for the next frontier. It was not unusual, though, when one of these older parents died, for the survivor to come and join some of the pioneer children in the new land.

The Helen valley pioneers did not come to start entirely new lives, for all of them came as families. The parents of these migrating families were older than might be expected, most in their thirties and forties. Local records show that many families migrated in groups, perhaps travelling with some of their neighbors and very often in the company of relatives.

Eight families acquired property in the Helen valley during the decade after it was opened for settlement. Half moved on in a surprisingly short time, leaving four who remained for many years. During the 1830s, the valley must have been a happy and energetic place. As the newcomers worked to build up their farms, more than 40 playful children ran along the banks of the Chattahoochee when not helping with the chores. Over the next several decades, the number of kids declined as families matured and the offspring moved away. However, there were still plenty of children in the valley: the census taker found only five slaves in 1840, but by 1860 there were 19 blacks, 14 of whom were children.

The four families who stayed on in the valley were the Bells, Pitners, Englands and Conleys. The average mother in these families had 8 children, birthing the first soon after she married in her early twenties and having the last at about age 42. Each mom had at least three children after coming to the valley. Although all of these women survived the maternal experience, the generations before and after were not so lucky, for the mother of Montgomery Bell and the wife of one of Adam Pitner's sons died at childbirth.

THE RESTLESS EARLY YEARS

Along with all of the Chattahoochee headwaters east of the Chestatee River, the land where Helen sits today was distributed in the 1820 Land Lottery, the third such lottery to be held in Georgia. Georgia had gone to the lottery system after massive frauds in the earlier "head right" system resulted in the sale of land grants amounting to an area estimated at three times the size of the entire state. Under the lottery system, as new land was acquired from the Indians, it was first surveyed into large districts, which were then subdivided into individual "land lots".

The size of the land lots varied according to the agricultural quality of the district. Lots in and below the Helen valley were a sizeable 250 acres, while those in the rugged mountains above were considerably larger at 490 acres. Generally, those eligible for draws in the lottery were males 18 and older who had been Georgia citizens for at least three years, although veterans of the Revolution, the War of 1812 and the Creek Indian War were also eligible even if they weren't current Georgians. Women could draw only as widowed survivors of eligible males; orphans could draw on the same basis. Entrants in all of these categories won lots on the upper Chattahoochee.

The Helen valley is in Land Lots 38, 39, and 40 in the Third District of what was originally Habersham County. None of the lottery winners ever lived on the lots, and it's hard to find any example nearby where a lucky drawer moved in to occupy his newly acquired land.[1] Whatever the State may have intended, the lottery seems to have made little difference in the long-standing mountain and piedmont migration patterns which played out across the Chattahoochee headwaters.

However, the new settlers did have one problem: they had to somehow acquire title from the lottery winners. The initial prices paid for the lots varied widely, ranging from as little as $20 to as much as $1500, with most selling for $150 to $300. As would be expected, the better lands often commanded a higher price. However, it appears that many lottery winners never even saw their prizes, and in these cases the prices had something to do with the actual property only if the buyer had seen it.

The three land lots in the Helen valley brought much higher than average prices. The Unicoi Turnpike crossed all of these 250 acre lots and each contained good bottom lands on both sides of the Chattahoochee River. By 1830, various members of the England family had owned all three of them. If they had all stayed, Helen might today be known as "Englandville", but in restless pioneer spirit, most of the Englands moved on after only a short stay in the valley. The land lots are shown on the maps as they were in pioneer days and as they are now. Starting at lower end of the Helen valley, here is a

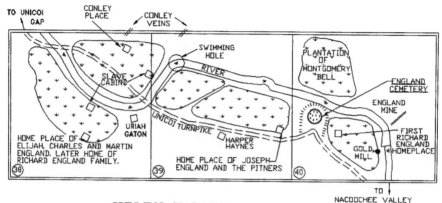

HELEN VALLEY, 1820–1840

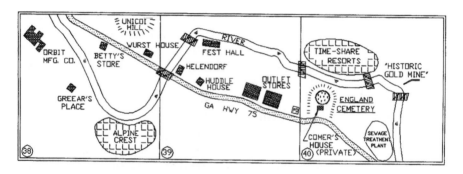

HELEN, GEORGIA, CIRCA 1990'S

ORIGINAL LAND LOTS. Georgia surveyors divided the Helen valley into three 250 acre Land Lots, which were distributed in the 1820 Land Lottery. After a restless early period when many of the first settlers chose to follow the departing frontier on to the west, four pioneer families remained for many years.

look at the first happenings on each of them:

Land Lot 40: This lot was won by Nancy Wood, a widow who lived in Elbert County. Although she won it in the 1820 lottery, she did not exercise her claim for five years, waiting until the last day of November, 1825 to pay the $18 dollar grant fee required by the State. Such a lapse was not unusual, for many lots went unclaimed for longer periods of time and quite a few — typically those least suited for farming — were never claimed. Mrs. Wood's belated decision to exercise her grant was evidently prompted by a purchase offer: only a month after obtaining title, she sold the lot to Mercer Fain for $600.

Mercer Fain was up to something, though, for he also sold the lot only a month after acquiring it, and at a considerable profit. Mercer had been in the area for years. His sister lived about two miles upriver from Lot 40 where the Robertstown community now is and his parents lived in the nearby town of Clarkesville, the county seat. His father was Ebenezer Fain, a tough old Indian fighter and Revolutionary veteran who had been a public servant in several of the many places he had lived.[2] Ebenezer served a few other things as well, for in 1823 he was found guilty of "keeping a tippling house open on the Sabbath Day".

Mercer may have actually lived on Land Lot 40 before he sold it, but it appears more likely that, upon learning of his buyer's interest and perhaps even his residence there, Mercer simply applied some of his father's savvy and got to the lottery winner first. In any event, Mercer Fain lived elsewhere in the county for several years after the sale, but by 1830 he had moved to Tennessee, continuing from there to Missouri and finally to Texas. Although Mercer left, other Fains eventually came to the Helen valley. Today, descendants of Ebenezer Fain can be found running Betty Fain's Country Store and the Nora Mills Granary. Although the Sunday sale of alcohol has recently been allowed in Helen restaurants, neither of these modern Fain establishments has any direct involvement with "tippling".

It was Richard England who bought Land Lot 40 from Mercer in February, 1826 for the hefty sum of $1500, giving Mercer a nice $900 profit for his month of ownership. Richard and wife Martha came from Burke County, North Carolina along with their five young children to occupy the bottomlands where the Helen sewage treatment plant and Comfort Inn are today. Richard paid a high price for his land, but his investment soon paid off more than anyone ever would have guessed. The next year, in 1827, he got half of his money back by selling some of his lands across the river to Montgomery Bell for $750. And by 1829, the England fields had proved to be good ones, generating enough corn so that Richard could spare 55 bushels to trade for 40 acres on an adjoining Land Lot.

The biggest payoff, though, was just around the corner. After gold was discovered, Richard England sold the part of his land where the "Helen Historic Goldmine" is today, recovering many times what he had originally paid Mercer Fain. Several years later, Richard England used some of his new found wealth to move his family to a spot he liked better. In about 1834, the England family moved from the lower end of the Helen valley to the upper end, where Richard acquired 125 acres on Land Lot 38. Richard still owned much of his original farm, though, including the small cemetery which had been established on the ridge overlooking his first fields and those of Montgomery Bell. Both the Englands and the Bells remained in the valley for many years. Their family stories are told in later sections.

Land Lot 39: Continuing up the Helen valley, this middle Land Lot was also won by a widow and then acquired by local men who soon sold it at a considerable profit. Three years after the lottery, Henry and Benjamin Crumley secured the lot from Sarah Adams for $500, giving her a mortgage note in lieu of cash. The Crumleys then turned around and sold the lot to Thomas Moore for $1000. Since Moore gave Habersham as his county of residence on the deed, he may well have been living on the lot and thus been forced into a high price in order to stay there and hold on to the improvements he had made.

Although goldminers and modern Alpine investors have done it many times since, Thomas Moore has the distinction of being the first person in the Helen valley to go bankrupt. In 1825, only two years after he acquired it, Moore's land was sold by the Habersham County Sheriff at public auction for $301. The purchaser was Devereau Jarratt, a rich man who operated the roadhouse where the Unicoi Turnpike met the Tugaloo River, a spot known as Traveler's Rest and now operated as a State Historic Site.[3] Four years later, in 1829, Jarratt sold the lot for $800 to one of his neighbors on the Tugaloo River. The neighbor was Joseph England, who subsequently moved to the Helen valley. Joseph came with some money, for before he left the Tugaloo, he sold his lands there to Jarratt for $2300.

Joseph was not among strangers when he settled in the valley, for members of the England family owned property on both sides. It's difficult to sort out all of the relations between the Englands since there were a number of them in the region — even some who had taken Cherokee wives — and branches of the family extending into South and North Carolina shared first names as well as the last. However, Joseph clearly appears to have been the son of the Charles England who was already living at the upper end of Helen valley on Land Lot 38, and was probably a more distant relation to the Richard England down on Land Lot 40.

Joseph was around 50 years of age when he and wife Matilda came to the valley. They brought seven children; two were under 10, four were teenagers, and the oldest twenty-something. Joseph's stay in the valley turned out to be only about two years. On April 9, 1831, ". . .being sick and like to die. . . but knowing that it is appointed only to die and being of sound mind and memory. . .", Joseph prepared his will. Neighbors Richard England and Montgomery Bell were witnesses. Upon his death shortly thereafter, Joseph in all probability became an early — and quite possibly the first —occupant of the England Cemetery, which sat upon the ridge only a few hundred yards from his house.

Joseph's will is one of the few from the period recorded at the Habersham County Courthouse.[4] To all of his children he gave a share of monetary notes owed to him. The rest of his estate he left to ". . .my beloved wife Matilda. . ." except for one horse, a yoke of steers, his two negroes, and two notes from Adam Pitner, all of which were to be used to pay his just debts. Among the possessions left to Matilda was a Land Lot in Coweta County, located in the lower Georgia piedmont near the Alabama line.

Joseph's family remained in the area for a short time after his passing, at least long enough for his orphan sons to draw in the 1832 land lotteries, the last to be held in Georgia. The boys did win a land lot, but like most lottery winners, they did not move there. Instead, Matilda soon left the Helen valley and moved to the Coweta County lot, accompanied by her dependent children. Things seem to have gone well in Coweta, for her son Powell married there in 1836 and "Matildey" was present and accounted for in the 1840 census. The move to Coweta also seems to have been planned well in advance, for Joseph had sold his Helen valley property even before he made out his will.

The two notes from Adam Pitner were the result of Joseph England's last major transaction, the sale of Land Lot 39. Joseph tripled his money when Pitner promised to pay $2500 for the property, which had by then been dis- covered to be in the Georgia gold belt. Coming from Tennessee, Adam Pitner moved in with his family, wife Rachel, and his father, John Pitner. The Pitners remained on this lot for the next fifty years, and are discussed further in a later section.

Land Lot 38: Stretching from the foot of Hamby Mountain and the Orbit manufacturing plant across the river to take in the Unicoi Hill and the original downtown area of Helen, this Land Lot was owned at least in part by four different Englands by the early 1830s. Like the other two Lots in the Helen valley, it was first acquired by a local man, but in this case he was the one who lived there. He was Elijah England, who bought it straight from the lottery winner in February of 1822 for the hefty price of $1000.

It was, according to the deed, the place where "said Elijah England now lives", so he had been building and clearing there for some time before the deed was signed. Since this is the earliest deed in the Helen valley, Elijah's was probably the first family to settle there. Elijah was about 30 years old when he and wife Elizabeth arrived with their three girls and one boy, all under age 11. William England, an older man who was probably Elijah's father, also lived with them. These Englands had lived in neighboring Franklin County before coming to the valley.

Although Elijah had occupied a favorable spot, something led him to leave the valley. One factor may have been the loss of his wife Elizabeth, for otherwise it's difficult to explain a curious entry in the Habersham County deed book captioned, "Deed of Gift of a Stud Horse":

> . . . to Sally Jane Moore of the same place [apparently the daughter of Thomas Moore who had lived next door on Land Lot 39]. . . I have given . . . a certain bay stud colt about 18 months old. . . . [signed] Elijah England, January 10, 1826.

There are no further records to indicate if this was a romantic interest or, if so, whether or not it was successful. Elijah did have a new spouse in his later years, although it was not Sally Jane. In any event, in 1824 Elijah sold half of his Land Lot for $725. In 1828, he sold the remaining half to Henry Highland Conley for $1000, so Elijah wound up with a good profit on his land even though he had paid a high price to get it. The Conleys were one of the four enduring pioneer families in the Helen valley and are discussed in a later section.

Elijah temporarily left Georgia, for when he executed a deed in 1832 for the sale of five negroes to Adam Pitner, he listed his residence as North Carolina. He soon was back, though, appearing in an 1834 census of Union County, living about thirty miles from the Helen valley, across the Blue Ridge near the present day town of Blairsville. Although this area was still in the Cherokee Nation, Georgia was pushing hard for their removal and had distributed the lands there by lottery. With the encouragement of the State, settlers had begun moving in, sometimes evicting the dispirited Cherokees to seize desirable sites.

In the 1860 census, Elijah was still in Union County, by then 70 years old and living with a 50 year old wife named Caroline. Slavery was never widespread on the back side of the Blue Ridge, but Elijah was one of 33 slave owners in Union, owning 6 of that county's 116 slaves.

When Elijah sold the first half of his Helen valley homeplace in 1824, it was purchased by Charles England, an older man who was a Revolutionary War veteran. In his earlier years, Charles had come from South Carolina to Georgia's Franklin County, where he lived on the Tugaloo River near his son Joseph England. As already mentioned, Joseph eventually followed his father to the Helen valley to live down on Land Lot 39.

Charles kept his half of Land Lot 38 for six years before selling it to yet another England in 1830. However, Charles was still around in 1832, when he drew for land and won in that year's land lotteries. From census records, the old soldier was at least 77 years old by then, and seems a good candidate for the England Cemetery if he didn't move on to Coweta County with his daughter-in-law Matilda and her youngest children.

When Charles sold his half of the lot in 1830, the buyer was Martin England, a son of Joseph England and therefore Charles' grandson. Martin was another of the restless Englands, though, for by 1834 he had migrated northward along the Unicoi Turnpike to live with his family of eight near what is now the town of Hiawassee. Like Elijah England, Martin had gone to the area which Georgia had declared to be Union County, but which was still in the bounds of the Cherokee Nation. Before he left the valley, Martin sold his Land Lot 38 property to Richard England, who became the fourth England to own it. Richard's move from the lower end of the Helen valley to the upper end was the shortest of the recent England migrations, leaving his family as the only Englands remaining of all who had come there in the early years.

When the Richard England family occupied the old home of Elijah, Charles, and Martin England, the last piece of a more permanent arrangement was in place. The frontier had passed through the small valley like a wave, leaving turbulence in its wake and carrying many pioneers forward with its momentum, but something more enduring was left when the commotion subsided.

A latter day picture of Martin England, taken some years after this pioneer followed the Unicoi Road across the mountain to settle near present day Hiawassee. His father Joseph England was probably the first of the Englands to be buried in the England Cemetery.

The family of Richard England would stay for a while, as would the Conleys across the river and the Pitners and the Bells down the way. There was traffic on the Unicoi Road and the Gold Rush caused a stir, but the valley was now a place where some would choose to live out their days. And time would inevitably add to the small cemeteries already established, as the old pioneers completed their earthly turn to become pioneers once again for those who remained behind.

THE PERMANENT SETTLERS

The Bells. The "plantation of Montgomery Bell" was located across the river where the time-share resorts at the lower end of Helen are today. The Bell property also took in a good portion of what used to be known as Buckhorn Mountain. A mid-sized stream arising on the mountain crossed the Bell place, and still bears the family name. Comer Vandiver says ginseng was once plentiful along this creek. Although it's called "Bell Branch" today, the stream is referred to in old deeds as "Bell's Mill Branch", indicating that Montgomery operated a grist mill there. There was also gold on the Bell place, for the bottoms where Montgomery grew his crops were worked for the precious metal during the 1830s.[5] Although deeds refer to his "plantation", the word in those days was used interchangeably with "farm", a description much better suited to the Bell Place.

The Bells place themselves perfectly in the Scotch-Irish migration, the name originating in Scotland and the descendants moving through Ireland, Pennsylvania, and Virginia before reaching the North Carolina foothills where Montgomery Bell was born in about 1785.[6] Family records say Montgomery Bell was the youngest of seven children and that his mother died at his birth. His father later remarried and moved to Buncombe County, North Carolina where Montgomery grew up and found his wife Comfort Brittian, a woman about 12 years his junior. The Bells came to America as Presbyterians, but many became Methodists in North Carolina and many local Bell descendants continue in that faith today.[7]

The Bell's religious experience is in keeping with that of their fellow settlers on the Chattahoochee headwaters. Although the Delaware River Valley group started out primarily as Presbyterians and Lutherans and the Chesapeake Bay group as Episcopalians, none of these churches made it to the Helen/Nacoochee area with the pioneers. By the late 1700s, American events — especially a religious happening known as the "Great Awakening" — brought Methodists and Baptists to the fore, particularly along the frontier, so that these were the only faiths represented in any organized fashion among the first settlers. Some members of the Pitner family and the last of the pioneer Englands are buried in nearby Methodist cemeteries, as are the members of Helen's other pioneer family, the Conleys.

Looking backwards along the migration paths, there were Presbyterian churches in neighboring Franklin County and also in the North Carolina highlands, where a few Lutheran congregations could also be found. From its English origins in 1729, the Methodist faith grew with amazing speed. The first Methodists to preach in upper North Carolina did not appear until the late 1780s,[8] but only a generation later the Methodists marched nearly alone with the Baptists in the subsequent migrations which swept across north Georgia and on to the west.

Although Montgomery Bell identified himself as a resident of Habersham County when purchasing his property from Richard England in 1827, it's hard to say when he moved there, for his daughter Louisa was born in that same year — but in North Carolina. The Bells do appear in the 1830 census of Habersham County with five children. Family records say Comfort and Montgomery had eight named children; the last two were boys born in the Helen valley. The 1840 census indicates that seven were still at home in that year, the youngest a two-year-old boy while the oldest were two girls in their twenties.

Montgomery had not been all that young when he came to the valley — he was in his early 40s — but his advancing years did not keep him from moving once again, for the Bells left the Helen valley during the 1840s. In 1842, Montgomery sold his property to goldminer A. J. Moses for $800, a sum 50 dollars more than he had paid for it fifteen years before. He and Comfort set out along the Unicoi Road to settle near Murphy, NC, which is about as far west as one can go in that state. There, an 1850 census taker found 65 year old Montgomery and 52 year old Comfort living with three teenage boys: Montgomery Jr., James, and William.

The Bells still had connections in Georgia, for daughter Eliza had married a Mr. Castleberry and remained in the county. By 1860, James and William Bell had returned to Georgia, showing up about five miles southeast of the Helen valley on property they acquired from the Castleberry family. James and William bought 100 acres near Yonah Mountain where they listed their occupation as "farmer". In 1860, both men were in their early twenties, married, and each had a newborn in the house. Between them, Eliza, James and William had 30 of the grandchildren of Comfort and Montgomery Bell, so that the Bells have many descendants in the county today.

Although Montgomery apparently died before 1860, the exact date and location of the event are not known. Comfort Bell lived to be 77. In 1873, she was buried in Chamblee, GA, where she had been staying with her daughter Louisa Bell Walker.[9] Chamblee was a country town at the time, but today has its own station on Atlanta's rapid rail system.

If Vic Bristol's clue is correct and there are Bells in the England Cemetery, it's hard to say who they are. With his move to North Carolina, it appears unlikely that Montgomery is there, although the possibility cannot be entirely ruled out since some of the Bells did return to the area. As reflected in the saying that "you don't count them until they're two", infant mortality rates were high in pioneer times. Family history says there were two Bell children — Carolina and Adaline — for which there are no records beyond their birth, so if there are any Bells in the old cemetery, they are probably these girls or other children who died young.

The Pitners.   As best the modern Pitners can figure out, their original American ancestor came from Germany to settle in Pennsylvania.[10] He was John Pitner, although his name was anglicized from the original spelling. It appears that John married and moved to Virginia, where he lived during the Revolution. There he had a number of children, the eldest being a boy which he also named John, born in 1763. After growing up and taking a wife, this second John moved down to eastern Tennessee near present day Knoxville by 1789, the year in which their son Adam was born.

Forty years later Adam Pitner became the sixth owner of Land Lot 39. The Pitner family almost certainly travelled the Unicoi Turnpike to reach the Helen valley, for their Tennessee home was at its northern end. John the second also accompanied Adam to the Helen valley, his wife having died sometime before 1830. These Pitners were Methodists, as Adam, his wife and father John are buried in the Methodist Cemetery in Nacoochee Valley.

The Pitner Place was beside the small stream which crosses under the main highway at the miniature golf course and runs past the Alpine Amusement Park on its way to merge with the Chattahoochee River. Indian artifacts and remains have been found along the lower reaches of this creek.

Few people know the stream has a name, a fact which testifies to its relative lack of importance today. But back when people didn't go very fast or very far in a day, and when traveling about meant splashing through the creeks along the way, even the smallest of streams was apt to have a name which was well known to the surrounding community. This little stream is sort of special, though, for it actually has two names left over from the old days, being known sometimes as "Pitner Branch" and on other occasions as "The Tanyard Branch". Both derive from its connection with the family of Adam Pitner, but the second takes a bit of explanation.

When Adam Pitner acquired Land Lot 39 from Joseph England in 1831, he didn't pay cash. Instead, he gave Joseph a mortgage note. The note was to have been paid by January of 1833, but was still unpaid ten months later. Two men had co-signed with Pitner, obligating themselves to pay up if he didn't. One was Edward Williams, who lived a short distance away in Nacoochee Valley.

As the months passed, Williams apparently became concerned and asked for protection against the possibility that Pitner might not pay his debt. In October of 1833, Pitner signed a "Mortgage on Tanyard Stocks" to give Williams ownership of "a certain tanyard and stock in said yard and tanyard tools and utencils. . . , said yard being on the plantation where said Pitner now lives" if Pitner didn't make good on the note.

This note shows how the small creek came to be known as the "Tanyard Branch" (with Tanyard pronounced as if it were two words: "Tan Yard"). Like it sounds, a tanyard was a place where animal hides were tanned, or made into leather. How the tanyard came to be there is part of a larger story which harks back along the Scotch-Irish migration path.

Long before the first of them came to the Delaware River Valley, the Scotch-Irish had been the traditional tanners and leather makers of Great Britain. They were even experienced with the hides of American white-tailed deer, which had been brought back on English ships for nearly a century before they started in earnest for the New World. And, for many years after their arrival, the Scotch-Irish remained closely associated with the hide trade, for several reasons. Simple necessity was first. Unlike the residents of the areas nearer the coast, the Scotch-Irish had to maintain leather making and other basic crafts in order to survive in their isolated highland homes. On the crest of the frontier, they at times even dressed in leather, wearing clothes of buckskin and hats of coon or mink, tails included.

Things were different down on the piedmont and coastal plains, where the focus was on raising cash crops. Large scale agriculture was the most profitable use of their time, generating profits with which to buy goods made elsewhere. As the flatlanders ignored traditional crafts like tanning to concentrate on their plantations, markets were created. For the mountaineers, the export of hides and leather became a source of much-needed cash.

Thus, in the 1830s, Adam Pitner's creek was known as the Tanyard Branch. There were other tanyards in the county, and a number of other north Georgia streams have also carried the name, including one which today is in downtown Atlanta. In the mid 1800s, the industrial revolution came to tanning. New processes and machinery eventually resulted in fewer and larger tanneries, leaving the old hand-operated ones to fade away. But even when the mountaineers lost their export market, they maintained their leather making skills for some years afterwards, for in their poor and isolated mountain region it was still a matter of necessity. In the Helen valley, another event along the Tanyard Branch would show that tanning remained a viable enterprise long after Adam Pitner finally paid his debt and canceled the "Mortgage on Tanyard Stocks".

Another feature of the Pitner place was a blacksmith shop.[11] The blacksmith appears to have been Harper Haynes, who lived on Land Lot 39 for a number of years. Harper first appeared in the official records in 1832 and was a witness on Richard England's will in 1835. The Haynes family was still in the Helen valley in 1840, although they disappear by 1850.[12] Haynes had a good location, for in addition to local business there was a steady stream of potential customers passing along the Unicoi Road. It's ironic to note there was still a blacksmith in the Helen valley in the 1930s, located on virtually the same spot Haynes had occupied a century before.

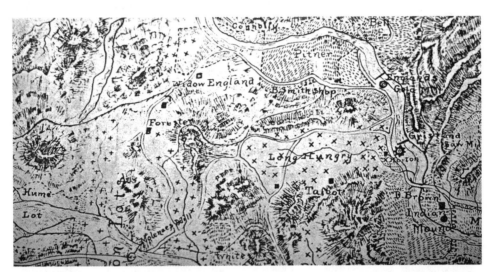

WINDOW TO THE PAST. This is an excerpt from a map entitled "Nacoochee 1837" drawn by R.C. Moffat, who signed it in 1881. After working as a local surveyor during pioneer times, Moffat moved away and years later drew the map from memory and whatever notes he may have had. The "Widow England" place is at the upper end of the Helen valley; "Englands Gold Mill" at the lower end is the site of today's "Gold Mines of Helen". Containing a wealth of information, the map shows a blacksmith shop on the Pitner place and Daniel Brown's grist and sash sawmill operation located where Nora Mills now stands. Mining sites are denoted by x's. Fields claimed from the Indians and the virgin forests are surrounded by dotted lines.

The only other thing known about Harper Haynes is that he may have had somewhat of a temper: he plead guilty to assault and battery in 1832. In this Harper was hardly alone, though, for court records from the period indicate rowdyism was not an unusual offense. Harper wasn't the first from the Helen valley to be so accused, either, since Joseph England had been cited for "affray" — officially defined as fighting in public "to the terror of the citizens" — in 1829. Assault and affray cases accounted for three-fourths of the nearly 400 criminal charges considered in Habersham County during the first thirty years after the county was formed in 1818.[13] And, as is still true of many such conflicts today, it's likely that most fights were settled by the parties themselves without the participation of the court.

The rough and ready spirit of the frontier was often encouraged by spirits of another kind. The 1840 census taker found 13 distilleries in Habersham County which had produced 4635 gallons of liquor in the previous year. An insight into the times was provided by an elderly pioneer who recalled the early days in a nearby mountain county:

> And what sort of folk were these people of the rude old days? Better than you have now. They would fight, knock down, and drag out, go home and get sobered up, come back and help you roll logs, build your house, run for the granny, wait on your sick, bury your dead, and mingle genuine tears with yours when sorrow came.[14]

Other than the episodes of assault and brawling, Helen valley residents were not formally accused of any crimes in the years before 1838. Elsewhere in the county, though, crime continued. Six murders were brought before the court. There were a few instances of mayhem, malicious mischief, and riot. Betting on horse and foot races, marksmanship, and wrestling matches was legal, but since gambling with cards and dice or on billiards was not, county residents were occasionally charged with "gaming".

Dalliances by married people led to eleven proceedings for adultery. As seems logical enough, usually the cases came in pairs, for only two men and one woman were charged alone. The first person to be accused of adultery was Garnette Smothers; when he was found not guilty, charges of fornication were dropped against Miss Milley Clark. Mr. Davis House was charged with adultery two years in a row. He was considered alone at first, but the second time around, Mariah Key was also charged. Both cases were "dismissed due to marriage", presumably that of Mr. House to the former Mrs. Key.

Since sexual misdeeds have always been pursued with greater discretion than transgressions such as fighting and assault, it's probably safe to assume that an even lower percentage wound up in court. However, Letitia Youngblood drew a little too much attention when she was charged with "keeping a lewd house". One case each of bastardy and bigamy along with three charges for "assault to rape" complete the list of sex related crimes which reached the Habersham County criminal docket.

Although only one case of bastardy appears, there were certainly many more instances of "fooling around", involving both young people who typically "did the right thing" if pa would allow, and married folks for whom "doin' right" was more problematic. Georgia law during the early 1800s required local officials to take action if they learned of an unwed mother giving birth. The "mother of the bastard child" was to be compelled to appear and either provide security for its support (funds the county would hold) or disclose the name of the father, who could then be made to pay a fine sufficient to support the child to age 14. The law looked good on paper, but like many moral works of the General Assembly, it seems in practice to have had less impact than intended. In the Georgia hills, illegitimates untouched by any official protection were common enough to have a nickname of their own: they were known as "woods colts".

Turning to other crimes, distilleries were legal, but selling liquor was still restricted on Sunday, leading to three cases for "keeping open a tippling house on the Sabbath Day" (including the one made against Ebenezer Fain). Four people were charged with the offense of "trading with a slave", several with "furnishing and selling spirituous liquors to slaves for their own use" and two with "negro stealing". Other reported thefts included seven instances of horse stealing and six of hog thieving. A handful of cases for cheating and swindling, forgery, counterfeiting, robbery, larceny, and burglary round out the first thirty years of criminal activity in old Habersham County, when the court considered about 14 cases a year.

Returning to the Helen valley and the Pitner Place, although his occupation was listed as "farmer" by the census taker, Adam Pitner didn't move south just to run a farm. Like many others, he seems to have been drawn to north Georgia by the allure of gold. With gold fever at its height, he soon sold interests in part of his big land lot to two Tennessee men for $2750. As was typical of the Georgia Gold Rush, these interests were traded further, eventually involving investors as far away as Savannah, Rhode Island, and New York City. And as was also common along the gold belt, much looking was done and a lot of money changed hands, but no major strikes were made.

During this period the gold deposits along Tanyard Branch were worked for the first time.[15] About ten years later, ownership of the lot was again held locally when the interests were sold at public auction on the courthouse steps. Part of the lot was then re-acquired by the Pitner family and the rest purchased by their neighbors the Conleys, who lived just up the River.

Adam Pitner had an even more direct connection to many events on the courthouse steps. For thirty-five years he appeared often in the deed records for Habersham County, and occasionally in what some considered the "misdeed" records as well. Adam eventually acquired interests in dozens of mining properties all along the Georgia gold belt. He apparently used some creative financing along the way, for a number of these properties were later sold at public auction. Many were purchased by his son Albert Galatin Pitner, who paid only a fraction of the price originally promised.

Pitner was also sued twice by his neighbors for non-payment of debts. Although Adam bought a negro woman and her four children from Elijah England in 1832, the census takers did not find any slaves on the Pitner Place in 1840 or 1850. Whatever might have transpired on his farm, Adam did use some blacks in his mining operations. One of the lawsuits against him was a "Debt for Rent" brought by a foul-tempered local slave owner named Moses Harshaw.

In July of 1844 Harshaw loaned Pitner six slaves, who toiled for the next nine months in Pitner's mines. Harshaw's complaint said he was to get one forth of the gold thus obtained and claimed that the mines "if properly worked" should have yielded 1000 pennyweights (50 ounces) of gold. Harshaw calculated that his share ought to be worth $225, which Adam Pitner had refused to pay even though "often requested to do so". But when the claim went to trial, it was apparently found that some of Harshaw's math was questionable. What "ought-to-have-been" didn't get much sympathy in the Georgia gold region where the locals realized that mines were more likely to fail than succeed, and the court ruled in Pitner's favor.

Having come to the Helen valley with five children, Adam and Rachel Pitner had two more soon after they arrived. By 1850, the two oldest had gone out on their own. Albert Galatin stayed in Habersham County for a while to put in his occasional appearances on the courthouse steps before finally moving off to the west Georgia town of Rome. John Pitner moved to Athens, GA where he opened a store that was later raided by General Sherman's marauding troops at the close of the Civil War.

John's store appears to have been successful from the start; by 1850 he had funds enough to buy the old Montgomery Bell Place and in 1855 he purchased the original homesite of Richard England, which included the England Cemetery. During the 1850s the third oldest son, Priar Pitner, acquired property far up on the mountainous headwaters of the Chattahoochee River, a location where few people have ever lived, but where Priar and his new wife Susan appear to have resided for a time.

By 1860, everything had changed on the Pitner place. Rachel had died in 1852 at the age of 59. She was buried in the nearby cemetery at the Nacoochee Methodist Church, in the family plot where Adam's father John had been buried ten years before. How long Adam and the remaining dependent children stayed in the Helen valley is unknown, but by 1860, Adam had remarried and moved off to Rabun County about thirty miles to the north. Adam did not forget his first wife or the place he used to live, though, for upon his death in 1873 at age 84, he was buried beside Rachel in Nacoochee Valley.

The 1860 census taker found the Pitner place occupied by a prodigal son, for 33-year-old Priar Pitner was then managing affairs beside the Tanyard Branch. Priar's first wife had died two years before when she was only 26, probably when giving birth to their only child, the then two-year-old Charles Pitner. Priar had remarried by 1860, the census showing his new wife to be a 17 year-old named Elizabeth. The new couple lived beside the Tanyard Branch for the next 20 years, birthing at least three more children along the way.

Priar Pitner died in 1880 at age 63, leaving Elizabeth and their two youngest children on the Pitner Place. Having lived and died on the same spot as Joseph England 50 years before, Priar likely made the same journey to the top of the ridge and his final resting place under the spreading beech tree in the England Cemetery. Priar is known to have died at home, and the cemetery at this time was owned by his brother John.

Although Vic Bristol indicated that "Pitners" were buried there, Priar is the only family member who can be identified. Any of Priar's children who died at home seem to be the only other suspects, for the rest of the Pitner clan is buried elsewhere. Priar may well have been the last person buried in the old cemetery. By 1880, the Bells and Englands had been gone from the valley for years and soon after Priar's passing his widow and children left, taking the Pitner name as well.

As was customary at the time, upon Priar's death three local men were appointed and an inventory made of his estate.[16] These listings of worldly effects give modern folks a tour of the farm and a glimpse into the homes of the pioneers. Priar's inventory included 28 items:

Part of Lot of Land No. 39
*   *and part of Lot of Land known as the Bell Lot*
*   *and lot No. 26 all known as the Home place*
                                   valued at   *$2000.00*
*   *1/2 the Jim Martin Lot No. 15/5th Dist valued at*   *20.00*

5 head sheep 1.50 each   *7.50*

3 cows and yearlings 15 [ea.]	45.00
5 head hogs 2.00 ea.	10.00
1 horse	10.00
1 buggy	20.00
1 wagon	20.00
plantation tools	10.00
10 sides of leather yet in vats	15.00
5 skins of goats and sheep yet in vats 1.25	
1 lot tanners tools	5.00
1 note on James Martin $35 dated Nov. 15th 1872	
with a credit [due] Dec. 10th 1874 $6.27	
2 sofas	2.00
1 sewing machine Singer	20.00
1 small table	1.50
1 whatnot [corner shelf]	1.50
1 book case and lot of books	5.00
1 doz. chairs	4.00
4 bed steads and bedding	50.00
2 buxxxrs [?]	7.00
2 wash stands	2.00
3 looking glasses	1.50
1 wash pot	4.00
1 stove	10.00
table furnature	5.00
3 large tables	3.00
1 armey musket	2.00
the present years crop	100.00
	$2382.25

The inventory reveals that, fifty years after Adam Pitner had signed his "Mortgage on Tanyard Stocks", there was still a tanyard on the little creek. Priar was only a boy when the old mortgage was signed, so he could not have been the original tanner there. He was probably an apprentice at the time, though, learning from his father and/or Grandfather John how to use the scrapers to remove the flesh and perhaps later strip the hair, which bark to use in the tanning troughs, when it was right to use alum, lard and flour or even brains to finish the leather, and how to string the skins on the racks to dry.

Priar and Elizabeth did have a few things the old folks didn't, though. The "whatnot" was a type of corner shelf of the Victorian style popular in the mid to late 1800s. Because it was of a later style than the type of furnishings brought by the pioneers, the whatnot indicates that the Pitners did not live in the cultural isolation which characterized some remote Appalachian areas. The sewing machine says much the same thing, for while they were a standard feature of middle class homes in the city, Mr. Singer's machines were a bit unusual in the mountains. These then-modern goods probably show the effects of the Unicoi Road, which brought a steady stream of outside influences through the Helen valley for over a century.

Something more can be said about the sewing machine. The day her new Singer arrived must have been a happy occasion for Elizabeth Pitner, for it made her life a lot easier. In fact, the sewing machine made such a difference in the lives of women that it's even been credited with helping foster the women's rights movements which began in the late 1800s. Although needlework remained a chore, by some estimates the sewing machine reduced a woman's work load by more than a third.

Elizabeth Pitner and the two youngest children stayed in the valley for a time after Priar died, but they were gone by the end of 1884 when she sold the Pitner place, including "the home where Priar Pitner lived at the time of his death". George Slaton next appears on the Tanyard Branch, arriving in 1884 and remaining there well into the 1900s. He ran a store and worked the Tanyard Branch for gold again before eventually selling most of his property to the Vandiver family, including the part which contained the England Cemetery. Mr. Slaton had known one of the pioneer Englands, an older man who lived where Unicoi Park is today, but the others had been gone from the Helen valley for many years before he arrived.

The Englands   Richard England was about 47 years old and wife Martha about 35 when they came to the Helen valley in 1826 or 1827. They brought five children under age 12 and had three more after they arrived. Richard's branch of the England family traces back along the Chesapeake Bay migration path, moving in three generations from Maryland to the piedmont region of North Carolina and then to the western foothills of that state before coming to the Helen valley.[17]

The Englands first settled in the lower end of the valley, but moved to the upper end in the early 1830s to occupy the large bottom area at the base of Hamby Mountain where Alpine Crest and Orbit Manufacturing Company are found today. By this time, Elijah, Charles and Martin England had all come and gone from this same spot. Although Richard and Martha would remain in the Helen valley, it would not be too long before many of their children did as they had done and headed off over the ridge to distant lands.

Richard England was not the only one of his immediate family to come to Georgia, for his mother and two siblings also moved to the area. His brother Elisha seems to have been the first in the clan to make the trek to Georgia, coming in the early 1820s to settle on Mossy Creek, a Chattahoochee tributary located below Yonah Mountain about 10 miles south of the Helen Valley. Elisha had several chances to make a second move in the frontier direction when he drew land in both the 1827 and 1832 land lotteries, but he chose to remain on Mossy Creek.

After Richard and Elisha's father died in North Carolina, their mother Margaret moved down to Mossy Creek to live with Elisha. She was there in 1827 when she drew in that year's land lottery as the widow of a Revolutionary veteran. She was still there five years later, when she got lucky and won a Land Lot in the 1832 Gold Lottery. This senior woman of the frontier did not go to her lot in the Georgia gold belt, but she did have another move left in her. Mother Margaret picked up once again and moved over the Blue Ridge to Union County, where she went to live with another of her descendants. She died there in the 1840s, becoming the first to be buried in the Old Choestoe Cemetery, where her weathered soapstone marker still stands.

Richard's sister Nancy lived a few miles away in Sautee Valley, apparently in a state of misery after having suffered the misfortune of marrying Moses Harshaw. All descriptions of Mr. Harshaw begin with the word "mean". Among other things, he is said to have killed his slaves when they got too old to be profitable, forcing them to leap from the cliffs of nearby Lynch Mountain or to dig their own graves if they were still able. When traveling about in his buggy, Harshaw sometimes had a slave in tow, the unfortunate black pulled along behind at the end of a long leather strap securely attached to a collar fitted snugly around his or her neck.

Many of Harshaw's other white neighbors had trouble with him as well; one was Adam Pitner, whom Moses sued over the aforementioned case of the slaves in the gold mine. Harshaw appears frequently in the early court records of Habersham County. At various times, he was found guilty of obstructing the legal process, of assault and battery, and charged with "conveying a slave convicted of a capital offense".[18]

Moses' lack of popularity with his many slaves showed up in the 1840 census, when two were "fugitives from the state" and he was the only slave owner to have any such property in this category. Hatred of Harshaw seems to have followed him even after his death in the early 1860s, for it's often said locally that his tombstone bore the words "Died and Gone to Hell". If so, it seems someone decided the epitaph was inappropriate, for the monument has long since disappeared and the grave is unmarked. A more colorful local tale has it that, when the devil came to get Moses, he just grabbed the monument as well.

Moses Harshaw's temperament inspired what at the time was a very rare occurrence, for Nancy England Harshaw sued for divorce. She was assisted in her petition by brother Elisha, and apparently by at least one of her sons. In 1843, Moses signed a bond giving Nancy the house in Clarkesville to which she had already moved, 4 negroes, some household goods, 1 bay horse, two cows and calves, and provisions and firewood for one year. All this was to be held in trust by their son Melvin on Nancy's behalf until the "Bill for a divorce from the claim and tie of matrimony" was settled. The proceeding took 7 years, but in 1850, Nancy finally succeeded in severing the claims and ties of Moses Harshaw.

Returning to the Helen valley, Richard England enjoyed his new homeplace at the upper end of the valley for only a short time. He was about 55 years old in 1835. As the year progressed, he became sick. In June, "being low in health of body but in perfect mind and memory", he made his will.[19] Richard still owned the part of his original lands which included the England Cemetery and shortly thereafter, this became his final resting place. He left the estate in Martha's hands until such time as all the children came of age or she married again. Martha never remarried, though, and the England plantation became known as "The Widow England Place".

The Englands always had company on their farm. They were slave owners, a practice consistent with the Chesapeake Bay culture. They had only one slave in 1830, but there were six living in the slave cabin by 1860. The England Place for a time was also home to Uriah Gaten, whose name still attaches to the small creek which flows through the Alpine Crest time sharing operation. Uriah's first appearance in county records was in 1822, when he was identified as insolvent and given an allotment of $56.50. But by 1840 Uriah was doing better. The census taker found his family of 11 prospering on the Widow England Place, with two of them applying themselves to farming and two working as gold miners.

By 1850, the three oldest of the eight England children had left home for good; all three lived across the Blue Ridge in Union County at the time of that year's census. In the 1840s, second-born son Daniel England built a home near what is now Vogel State Park. Located on US 19 & 129, this home is unoccupied but still standing.

Something about Union County made it a very popular England destination, for a lot of them settled there. The families of Elijah and Martin from the Helen valley, one of Elisha's sons from Mossy Creek, and even Richard England's mother — all of these had preceded this latest group of England migrants to Union County. By 1860, Martha had 13 grandchildren living in Union County, and there are many descendants of Richard and Martha England living there today.

THE WILL OF RICHARD ENGLAND

In the name of God amen. I Richard England being low in health of body but in perfect mind and memory do make this my last will and testament, viz, first I will that all my just debts be paid (second) I will and bequeath to my well beloved wife Patsey England to have the rest of my estate after my debt being paid until my youngest child comes of age, then I will that my estate be equally divided among my children, the widow to take a child's part and if my wife should remarry before the youngest child comes of age I will that all my estate be divided among my children and further I will that my wife and Athen my oldest son to execute my will and I further will that if any of my oldest children should marry before my youngest child comes of age that they should have half of their supposed part to be accounted for by them at the divide and if they should refuse to account they shall then have no more. If my wife should marry before my youngest child is of age I then will that my estate be sold and equally divided Athan and Daniel to have the control of all the non-age shares until they become of age by giving bond of security for the prompt payment unto them when they come of age given under my hand and seal this 5th day of June 1835.

WITNESSES

Robert Westmoreland
Harper Haynes
Comfort Bell

In simple fashion, Richard England's will provided that his estate be used to support his wife and minor children for as long as they needed it. After the salutation, the will is only two sentences long. Mrs. England's given name was "Martha" but Richard called her "Patsey", a nickname.

In their travels across the Blue Ridge, the Englands followed paths established by Indians and migrating animals thousands of years before. The ancient Choestoe Trail branched off the Unicoi Road about three miles above the Helen valley, following Low Gap Creek and Trail Ridge to cross the mountain at Low Gap. To the west, the Tesnatee Trail crossed at the gap of the same name.[20] Although the Choestoe Trail largely remained in its primitive state, the Tesnatee route was improved and in operation as the Logan Turnpike by the 1830s. Today, although it takes a different route up the mountain, the Richard Russell Scenic Highway generally follows the old Tesnatee route as it descends into Union County.

By the time of the 1860 census, she was long a widow and now a grandmother many times over, but 68-year-old Martha England was still at home in the Helen valley. Living with her were middle sons Jerome and Coleman, both in their mid-30s, and her youngest child Mary Ann, who was 29 years old. Six blacks lived in the slave cabin, a 32-year-old female and five children under age 15. The England family had been in the valley for nearly thirty-five years, but many changes were just around the corner for the remaining residents on the Widow England Place.

Many children of Richard England and Martha England migrated over the Blue Ridge to Union County, following ancient paths like the Choestoe and Tesnatee Trails. This couple is said to be second-son Daniel England and his wife (who appears to be pregnant), photographed about 1850.

Both Jerome and Coleman England served in the Civil War. Coleman returned home on sick leave and died there in October, 1863. County tax records indicate that mother Martha died at about the same time. Both could only have been buried with Richard in the England Cemetery, and Coleman and his mother were probably the last of the family to be buried there. Mary Ann finally married and moved off to Union County, joining her three oldest siblings in that most popular of England destinations across the Blue Ridge. With Mary Ann's departure, only Jerome remained in the Helen valley.

Upon Coleman's death, the usual estate inventory was performed. Coleman was half owner of the estate; the other half was Jerome's. The listing was as follows:

Inventory of Coalman England Estate, third day of February 1864.
Whitheth is one undivided half of the following property to wit

	Land one half lot No. 38 in the 3rd dist.	*600.00*
1	*Negro woman Caroline 35 years old*	*300.00*
1	*Negro boy Frank 18 years old*	*500.00*
1	*Negro girl Jane 12 years old*	*250.00*
1	*Negro boy Sam 10 years old*	*250.00*
1	*Negro boy Henry 8 years old*	*200.00*
1	*Negro girl Josephine 6 years old*	*150.00*
1	*gray mare 21 blind*	*6.50*
1	*bay horse 12 years old*	*22.50*
1	*gray horse coalt 3 years old*	*20.00*
1	*parte red cow & calf 13 years old*	*5.00*
2	*yearlings 3 years old*	*8.00*
27	*head of hogs, sows, shoats, pigs*	*54.00*
2	*head of sheep*	*7.00*
	15.5 Bushels of corn at one dollar per bushel	*15.50*
	10 fodder shucks and straw	*10.00*
	2 sythes & cradles	*4.00*
	100 pounds of old wagon iron	*3.00*
	1 patent plow	*3.00*
	farming implements in all	*16.00*
	8 bushels of rye	*4.00*
	50 pounds bacon	*31.25*
	1 loom	*4.00*
	1 iron pot rack	*.75*
	pot ware & cucking utenticls	*3.00*
	1 grind stone	*2.00*
	1 pare of stylyards	*1.00*
	1 clock	*5.00*
	1 set of dog irons	*1.00*
	1 side board	*1.50*
	2 tables	*1.00*
	3 bedsleds and bedes & beding	*30.00*
	6 chairs	*1.50*
	50 pounds of lard	*3.12 1/2*
	20 pounds of tallow	*1.00*
	214 lbs. of xixs grain [?]	*6.67 1/2*
	3 bells	*.93 1/2*
	lether in all	*2.50*
	103 cash in hand	*51.50*
	49.10 accounts on D.M. Horton	*24.55*
	87.90 accounts on Wm. Lawrence	*43.95*
		2644.73 1/2

Accounting for over 60% of the appraised value, the slaves were easily the most valuable part of the England estate. But about a year after the inventory was taken, the Civil War was over and they were worth nothing to their former owners. The blacks then left the Helen valley. Another item of interest is the loom. Martha was shown in the 1840 census as being profitably engaged in a craft of some sort; the presence of the loom and the female slaves indicates she may well have made cloth to sell. Although there were only 2 sheep at the time of the inventory, the county tax man had recorded about 10 a few years earlier, enough for a good supply of wool.

EARLY LOOM. This loom was employed for a century and a half. After years of service in North Carolina, the loom was brought to the Georgia mountains by early settlers, where it remained in use until 1940. One of the fabrics produced on the loom was "linsey-woolsey", which mixed linen threads for strength with woolen ones for warmth. Linen was made from locally grown flax. Dr. Tom Lumsden holds a "flax-hackle" used to separate flax fibers from the plant's woody stems. The loom and a growing collection of Indian, gold mining, and pioneer artifacts can be seen at the Historical Museum in the Sautee-Nacoochee Community Association Community Center on Highway 255 near the Old Sautee Store.

Jerome England was the last of the pioneer Englands to live in the Helen area. In 1867, he sold the England Place to J.R. Dean, a Massachusetts school teacher turned big-time goldminer whose family kept most of the property for the next 60 years. Jerome later purchased land just over the next ridge, along Smith Creek where Unicoi State Park is today. Jerome married and had two children, but both moved away in the late 1800s and neither had children of their own.

Jerome died in 1890 at the age of 68. He is buried at the Methodist church in Robertstown, the only member of the England family in the Helen area to be buried with an identifying marker. Jerome had been a trustee of this church when it was founded in 1860. In 1891, Jerome's children sold the

Smith Creek property to Sidney Smith Crumley, a Methodist preacher who later became the last gate-keeper on the Unicoi Road.

The Conleys The site of the Conley cabin is one of the prettiest spots in modern Helen, for it sat on top of what is today known as the Unicoi Hill. From this vantage point, the Conleys watched over fields and pastures stretching along the Chattahoochee River bottoms from the Feste Hall bridge nearly around to Betty Fain's Country Store. A good portion of this area was worked for gold in the 1830s.[21] There are also several tunnels on the old Conley Place, including a small one in the ridge overlooking the Unicoi Hill. The Unicoi Road crossed the Conley place about where Main Street runs today, fording the river at the Main Street bridge and again at the upper end of the valley.

The Conley family history says the original bearer of the name was a true Irishman who came to America in 1743.[22] He was John O'Connelly, who settled on the Catawba River in old Burke County, North Carolina. The introductory "O" was dropped long before Henry Highland Connelly was born three generations later. Henry was still going by "Connelly" when he married Nancy Anne Brown in 1822, but when he came to the Helen valley about six years later, he was using the modern form, signing the deed as "Conley". Henry's lyrical first names came from a prominent citizen of Burke County, who of course was Henry Highland.

Henry was about 30 years old and wife Nancy 27 when they acquired their Helen valley property from Elijah England in 1828. The Conleys brought three youngsters to their new spot on the hill and were soon busy having more, for Nancy eventually had 10 children during her 21 year child-bearing career, the last being born when was she was 46. Nancy was a daughter of Daniel Brown, who had earlier come from North Carolina to settle near the Nacoochee Indian mound.

This is a convenient point to look at some of the lore about the settlement of the Helen/Nacoochee area. Anyone delving into local pioneer history soon encounters a curious document known as the "List of 61". This writing purports to tell the story of the first settlers in two handy paragraphs before offering a list of the 61 original families in the area, all of whom supposedly came from North Carolina at the same time. The Conleys are the third family on the List; the Englands, Bells and Pitners are also included.

From its creation in 1922, the List has taken on a life of its own. A copy seems to reside in almost everyone's history drawer, and parts of it have even been cast in bronze on an official historical marker in Nacoochee Valley. Almost no one knows who wrote it, a fact which only serves to give it an embellishing mystical quality. As is typical of such tales, the List is built around a kernel of truth, but it is a great exaggeration and wildly inaccurate in its details. Some excerpts:

> There were 61 different families that came in two parties, one just one day behind the other, using the same camp fires, one party from Burke County, North Carolina and one from Rutherford County, North Carolina. . . they left. . . on March 1st, 1822, arriving at Nacoochee on the 10th. . . Travel was made more difficult by reason of the fact that there were no roads a good part of the way and they had to cut their way through forests. . . . They brought roses, some of which are still living. I have roses of four different bushes that were brought in that party. They also brought the Virginia and Johnson grass, etc. . . . They were the first white people in this county and bought land from the Indians at one cent an acre.[23]

The 61 families then follow, of which it was written some years later ". . .the list of members of those first wagon trains sounds like a cross section of people living there today. . ." Notice in this comment, the "parties" have now become "wagon trains". This shows something else about the list: different versions have arisen since everyone who recounts the tale tends to add their own twists to it, often coloring the story with popular notions of the day, which in this case appear to come from Hollywood westerns. And it's not surprising that the List was a pretty fair "cross section" of people living in the Helen/Nacoochee area at the time it was compiled and even for decades afterwards, since most of them were descendants of pioneers. Beyond that, though, the List gets off the track.

The true fact around which the List is built is that many local families did come from North Carolina, and there is little doubt that a group came from there at about the same time. However, many people on the List came from Georgia, South Carolina, and even Tennessee, and all of the North Carolinians did not come from the same two counties. Some of the North Carolinians were early arrivals, but the very first settlers in the area were Georgians like Elijah England, who had already moved in and obtained title to their land before the "wagon trains" are said to have rolled. And those on the List who did come from North Carolina did not all arrive at the same time. In the Helen valley, the Conleys, the Bells and the Englands are examples, since all were still having children in North Carolina years after the List places them in the Helen valley.

The List of 61 says that Daniel Brown purchased 2000 acres of land from the Indians at 1¢ per acre, including the entire Helen valley, which he subsequently gave to the Conleys. Variants of this tale are part of the pioneer lore of many areas, but they are without any basis in north Georgia. At the time, Cherokee law allowed the release of their lands only through treaties approved by their National Council. Anyone violating this prohibition was subject to death, a penalty which the Cherokees imposed on at least one occasion. The story was much the same on the American side, for no white citizen could obtain land directly from the Indians. And, of course, the Indians and the federal government had negotiated a treaty in 1819, giving the Chattahoochee headwaters to the State of Georgia at least three years before Daniel Brown moved to the area.

Since it took years to construct the Unicoi Turnpike through the rugged Appalachians, it's difficult to imagine the first party of North Carolinians hacking their way through virgin forests to accomplish a similar feat in only 10 days, even while leaving a clear path and a good supply of firewood and hot coals for that lazy bunch which followed in the second "wagon train". In fact, there had long been wagon roads running from the North Carolina foothills across South Carolina to the Savannah headwaters, and these are the routes the North Carolinians almost certainly used to reach Georgia.

But no matter how inaccurate the List of 61 might be, it is certain to endure. A simple and engaging tale with an authentic ring, it's the only locally available story of the frontier. The mythical List was created by a well-intentioned local historian who relied almost exclusively on word of mouth to compile it. However, a hundred years had passed since the first settlers arrived, leaving the frontier days well beyond the reach of living memory. Since the List derived from oral accounts, it was based on a century-long process like the childhood game of "Pass The Secret", with the local historian at the end of the line. In proper fashion, he added his touches to the story and passed it along for the addition of "wagon trains" and other embellishments in the years to come.

Returning to the Conley place, at the time of the 1840 census Henry and Nancy had eight children still at home, all under the age of 20. Two more youngsters were yet to come. Henry and father-in-law Daniel Brown were listed as being engaged in a commercial activity; Henry may well have had an interest in Brown's grist and "sash" sawmill enterprise. Located where Nora Mills is today, this water-powered milling operation supplied cut boards for local construction by 1830. The oldest known home in the area (the Richardson-Lumsden House on Dukes Creek, built in 1831 according to a date on the chimney) stands about a mile from the Brown sawmill site and is constructed of sawn boards produced from a similar mill.[24]

Sawn boards did not entirely replace logs for many years, though, and they were even used in log construction. In many north Georgia log cabins built during the early 1800's, the logs were not hewn for a tight fit nor was mortar or red clay "chinking" used to fill the spaces between the logs. Instead, logs were stacked with minimal notching and the gaps between them covered with horizontal "weatherboarding" to seal out the elements. Tapered weatherboards could be split or "rived" from logs in short sections, but longer cut boards were preferred where available. Once the log walls were constructed, the weatherboards were nailed along the squared-up faces of the logs, sometimes on the inside but more often on the outside. As they evolved, many cabins eventually received a full covering of weatherboarding, becoming virtually indistinguishable from frame structures.[25]

Although they had no slaves in 1830, there were three female slaves living on the Conley farm in 1840. By 1850, the number of blacks there had doubled. It appears the same

LOGS AND WEATHERBOARDING. This close-up of Daniel England's 1840s cabin shows the logs used in its original construction and some of the weatherboarding later used to cover them. The largest log measures over 20 inches across. Weatherboards were "rived" or split until sawn boards were available. When sawn boards were applied to the inside walls as well, the logs were almost completely hidden.

three females were there, accompanied by two baby boys and a 65 year old man. At the time, about 200 out of the over 1000 slaves in Habersham County lived in the Helen/Nacoochee area. Ten years later the number of slaves on the Conley place had doubled again, for the 1860 census taker found 12 blacks there, seven of whom were children under the age of 10. However, the 1870 census taker found a different situation.

The end of the Civil War brought confusion for both blacks and whites. Although they had been freed, facing the prospect with no land and little money seemed a bleak prospect to many blacks.[26] A number of former slaves made arrangements to remain in the Helen-Nacoochee area. Some of these took the last name of their former owners and stayed on to toil in the same fields for years; many of their descendants are around yet and living in the nearby Bean Creek community, just as much the sons and daughters of pioneer stock as many of their white neighbors.

In the Helen valley itself, though, the blacks living on the Conley farm and the England place across the river did not stay. The circumstances of their departure are not known, but they were gone by 1867 when they were missing from a listing of registered voters which the conquering Federals required all local governments to compile. Of the many blacks who left the area, some left on their own. Others were forced out, being dismissed by their former owners and told to fend for themselves.

If the Conley or England blacks remained nearby, they did not take either of these names. Judging from 1860 and 1870 census records, it's possible that at least some of the freed Conley slaves went to Atlanta. For each owner, the 1860 slave census listed slaves by sex and gave the age of each. Unfortunately, it did not record any slave names. There were no white Conleys in Atlanta or its county of Fulton in 1860. Ten years later, a number of blacks appeared in the 1870 census bearing the Conley name. Two with households in Atlanta were of the right age to have been on Henry Conley's list ten years before: a black male named Kenion Conley and a female named Jenny Conley.

The 1870 census also shows about a half dozen black families bearing the England name. The first names of the England slaves in the Helen valley are known since they were listed in Coleman England's estate inventory in 1864, but none of the blacks listed in 1870 match those on Coleman's list. However, on the back-side of the Blue Ridge, several blacks bearing the England name in 1870 did take it from former residents of the Helen valley.

In Union County, an 18-year-old black female named Billy and 26-year-old Alexander England apparently took their last names from Elijah England, the original owner of Land Lot 39 in the Helen valley. And over in Towns County near Hiawassee, a 13-year-old black female named Lewce England probably borrowed her new last name from Martin England, who had owned two of Towns County's 108 slaves in 1860 and also once lived on Land Lot 39. Although blacks have always lived in the Helen/Nacoochee area and in neighboring counties along the front of the Blue Ridge, all of the blacks eventually left Union and Towns counties.

Henry Conley mined his lands for gold after its discovery in 1828 or 1829, working the bottoms along the Chattahoochee and digging several tunnels, including a small one in the ridge above his house. Henry got an added touch of gold fever about 1845, when he joined with two other men who had also married daughters of Daniel Brown to acquire three nearby gold properties. What success they may have had is unknown, but like other locals Henry had seen enough to know the risks of mining, making it unlikely he bet too much on the venture. In any event, the mining interests stayed in the family, for nine years later the three men sold their stakes to one of Daniel Brown's sons.

By 1850, the two oldest Conley boys were in their twenties, and both had left home. By 1860, the three oldest daughters had done the same, leaving the five youngest children on the Conley place. Many went in the westward direction, three across the Blue Ridge to join the Englands in Union County and two others to northwest Georgia. By the 1870s, only two of the Conley children remained in the Helen valley. One was John Conley, who had appeared in the 1870 Union County census but returned to the Helen valley to build a house on the Conley lands across the river and remain there for the rest of his days. Although the Conley name is no longer around, they have many descendants nearby. Several of the Conley girls married locally and John Conley had many daughters who did the same.

The last to live on the original homeplace was Sam Conley, who remained there to care for his aging parents, both of whom lived 80 years. Henry Highland Conley died in 1878 and wife Nancy in 1881; they are buried together in the Nacoochee Methodist Cemetery. Sam remained on the Conley place for some years afterwards, but his story did not end there. Sam was the last of the pioneer children to leave the Helen valley, headed like most of the others for western lands. Sam left something the others did not, for he was a keeper of diaries who left a written record of his years there. Sam's diaries give a detailed look at life in the valley, and are recounted in the next section.

* * * * *

LEGACY

As the pioneer parents of the Helen valley grew older, they must have looked back on their accomplishments with some satisfaction. Things were hard at first when there were children yet to be born and fields yet to be cleared, but all of these families succeeded in their quest to prosper in a new land. Montgomery and Comfort Bell turned out to be the restless ones, moving on to North Carolina, but the others were largely content to remain on the Chattahoochee headwaters. Considering the times, none of them died young.

Richard England died at 55 and Rachel Pitner at 59, but the others lived from about 70 to as long as 84 years. The Conleys were the most enduring of the pioneer families. When Nancy Anne Conley died in 1881, she was the last of the pioneer parents.

Nancy's son John Conley was the last of the pioneer children in the valley, remaining there until his death in 1890. John's widow and some of his daughters remained in the valley for some years and there are many Conley descendants in the area today, although none carry the Conley name. Thanks to daughter Eliza and the two youngest boys, Montgomery and Comfort Bell also have many descendants in the county. The nearest Englands are about thirty miles away, over the Blue Ridge in that favorite England place, Union County. Pitner descendants are the hardest to find. Although there is a Pitner genealogy group and many recognize themselves as descendants of Adam Pitner, none appear to be nearby.

As these families dispersed from the Helen valley, memories of the early days went with them. No one came any longer to check on the England Cemetery, leaving the spreading beech tree to watch over the pioneers who remained behind. By the time the ubiquitous List of 61 was compiled in 1922, much had been forgotten. Although the four enduring pioneer families show up on the List, it contains errors on each of them. The local historian mistakenly changed both Montgomery Bell's and Richard England's first names to "William", moved Henry Conley's house across the river, and had Adam Pitner coming from North Carolina instead of his native Tennessee. And the other pioneers in the valley went undetected, for neither the families of Elijah, Martin and Joseph England nor that of Thomas Moore appear on the List.

As best they can be identified, the occupants of the England Cemetery are Joseph, Richard, Martha and Coleman England, and Priar Pitner. If Richard England and Priar Pitner had any children who did not survive, they may also be buried there. Charles England, the aged veteran of the Revolution and father of Joseph England, was still around in 1832 and could be buried there as well. The only Bells likely to be there are any children who died young while the Bells lived in the valley, although there is a slight possibility that Montgomery Bell could be buried there. Even though he moved off to North Carolina, the infamous List of 61 places him there, although under the above mentioned alias of "William Bell", which was in fact the name of his son who is buried in nearby Cleveland.

Over the years, other people lived on the property owned by the Englands and the Pitners, working as tenant farmers, as gold miners like Uriah Gaten or at various trades as did blacksmith Harper Haynes. Some of these may also be buried in the old cemetery, but with no clue to make the connection, it's unlikely their identities will ever be known.

PIONEER CABIN. Located along US 19/129 in Union County near Vogel State Park, this home was originally built by Daniel England in the 1840s. Like most old cabins, the addition of weatherboarding and a tin roof have changed its appearance. Daniel came to the Helen valley as a child in the 1820s. As did many of his brothers and sisters, he migrated over the Blue Ridge to obtain a place of his own.

THE EARLY YEARS

By the time Samuel Oswal Conley was born in the summer of 1844, the pioneer settlers had been living in the Helen area for over two decades. Sam was the seventh of eight Conley children, the fifth of six born in Georgia after the family moved down from the North Carolina highlands in the 1820s. Sam's younger sister Hattie (a nickname for Harriet) was the last member of the family, born three years later in the spring of 1847. Sam and his little sister seem to have been quite close. As events connected with Hattie would show, Sam was not a man to forget those dear to him.

By the 1840s, local farms were well-established and productive. Like other nearby landowners who worked the rich fields along the Chattahoochee, the Conleys owned slaves. They had only a few when Sam was born, but by the Civil War the two slave cabins on the Conley farm were home to a dozen blacks.[1]

Goldminers still labored in the hills as they would for years, although many had come and gone since the initial euphoria of the Georgia Gold Rush. Sam surely did some prospecting as a boy, for there were plenty of gold deposits on the Conley place.

After twenty years of occupation, some of the pioneer settlers had already found their final resting place, joining the remains of their Indian predecessors in what was now Georgia soil. Pioneer children had grown up, too, with many choosing to follow in the pattern of their ancestors and continue on the great American migration to the west. By the mid 1800s, Sam had a scattering of relatives in that direction, including relations in Gainesville and Atlanta, a brother living a day's ride across the Blue Ridge near Hiawassee and two sisters residing near the towns of Dalton and Chatsworth in the northwest corner of Georgia.

In the 1840s, young Sam could go a few miles down the river to Nacoochee Valley and there attend church, shop at a country store, mail a letter, and go to school. Northeast Georgians still looked eastward towards Augusta, Savannah, and Charleston for the affairs of commerce, but the settlement which would become Atlanta was already established and the focus would soon swing rapidly to the west.

It's not so noticeable from the modern perspective, but the

Chattahoochee River in Helen has changed over the years. Here and there its channel has been moved by goldminers and railroad men, and occasionally by raging floods, or "freshets" as they were once called. The laurels and rhododendrons have been stripped away and the river banks raised in most places to accommodate roads and buildings. But since the underlying rock is for the most part undisturbed, some features remain the same. The shallow fords where old roads once crossed are still there, even if the banks are now steep.

The rock formations which give rise to deep places are still there too. The deepest of these are used as swimming holes. Although floods have lately filled some with rocks and sand, there are several places which over the years have played host to rope swings and happy people who shared the icy waters with the trout for as long as they could stand it.

The best swimming hole around Helen has always been the one known as the "Big Rock", and it's on the old Conley place. Depending on what the floods do, it can be over ten feet deep. Summer evenings in the mountains are usually cool enough, but the days can be plenty hot and steamy. It's hard not to imagine Sam and his contemporaries having done what latter-day Helen children still do on hot afternoons, for summer relief has always been in the cool waters of the Chattahoochee.

Although many mountain children got little in the way of formal education, the Conley children attended school. Sam put his writing skills to use in a series of diaries which he kept for much of his adult life.[2] Most of the diaries are recorded in small booklets distributed for the purpose of advertising patent medicines, where Sam's matter-of-fact reporting appears on one page and cures for a host of often comical diseases on the next. Typically, the entries are brief notations on the notable events of the day or the week, and not meant to describe his inner feelings or tell the stories behind the events he records. However, they do allow an otherwise unattainable first-hand look into 1800s life in the Helen valley.

Sam's diaries begin in 1863 when he served as a foot soldier in the Confederate Army of Tennessee. He came home sick for a while, but rejoined the army just in time to fight against General Sherman at the battles of Kennesaw Mountain and Atlanta. Much of his account, though, chronicles the epic final journey of the Army of Tennessee after its expulsion from Atlanta.

After the war, Sam returned home to continue his schooling and work on the Conley farm. A late bloomer of sorts, Sam was the one who stayed on the homeplace and helped his parents in their later years. His diaries for a long time after the war are those of a single man, recording a lot of names and activities having to do with eligible mountain girls.

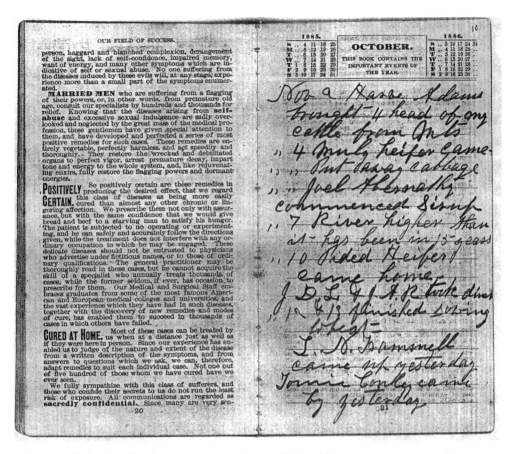

Many of Sam's diaries were kept in small booklets given out to advertise patent medicines.

The diaries reflect the rhythms of life on a mountain farm. They also document Sam's many travels across north Georgia, usually by horse but also via the rail lines which came to the area in the years after the war.

Finally, the happy conclusion to the romantic adventures of the single Sam Conley appears. There are no poems or sentimental lines in the diaries, but Sam's excitement over the future Mrs. Conley shows in more ways than one. Sam's marriage led first to some changes on the Conley place, then to a tribute of sorts for little-sister Hattie, and ultimately to his departure from the Helen valley.

A FOOT SOLDIER IN THE TENNESSEE ARMY

In many areas of the north Georgia mountains, there was considerable opposition to the Civil War. Since mountaineers typically worked their small farms themselves, they often had no interest in fighting to preserve slavery or abolish tariffs. On the back side of the Georgia Blue Ridge where there were few slaves, some counties preferred leaving Georgia to seceding from the Union, and there have always been Republicans across the mountain even in the century after the war when they were scarce elsewhere in the state.[3]

In the Helen/Nacoochee area, though, the sentiment was more in line with the rest of Dixie. The 1860 census listed 183 slaves in White County, and about a thousand in neighboring Habersham County. Most of the White County slaves were held by owners of the larger farms stretching along the Chattahoochee River from Nacoochee Valley to the Trammell Place (now Robertstown) about a mile upriver from the Conleys. Even though few local families had slaves, many young men from the Chattahoochee headwaters fought and did their share of dying for the Southern cause on battlefields from Gettysburg to the Mississippi River.

In February of 1863, following the lead of his older brothers, nineteen-year-old Sam Conley became a soldier in the Army of Tennessee. In the year that followed, he was hospitalized for illness at Loudon, Tennessee and then at Rome, Georgia before finally being sent home on 30 days sick leave in April of 1864.

Sam re-joined his infantry unit on May 30, just in time to help the 65,000-man Tennessee Army make a stand against General Sherman's invading force of 98,000 at the Battle of Kennesaw Mountain about 20 miles northwest of Atlanta. The outmanned confederates had been unable to stop the Union forces as they pushed southward from Chattanooga, but hoped to take advantage of the natural defenses afforded by the mountain to keep the Yankees from reaching Atlanta.

For the next seven weeks, the two huge armies maneuvered and occasionally fought against each other in the vicinity of the mountain. Rains were heavy during the first part of the engagement, leaving soldiers to struggle along slick roads where the mud was "from half a leg to knee-deep".

As the Generals plotted their moves, soldiers like Sam were constantly moved about like players on a chess board, their days filled with a mixture of tedium and hard work as they awaited the fights which were sure to come. Sam's diary records the activities of a foot soldier between battles: digging in, going out on advance picket lines, returning to the main ditches, moving to the rear to be held in reserve, being shifted along the line to counter enemy moves or try for tactical advantage.

The Confederates had selected the right place to make a stand, but Sherman did not make the mistake of wasting his forces against their superior defensive positions. Instead, he finally decided to bypass Kennesaw Mountain and continue his march towards Atlanta, forcing the Army of Tennessee to fall back and defend the city. Union forces crossed the Chattahoochee River in late July to begin the assault on Atlanta.

As fighting raged first north and then east of the city, Sam's division fought at Peachtree Creek and the town of Decatur. The defenders got the worst of these clashes, suffering 15,000 casualties versus just 6,000 for the Union forces — a very bad ratio for the defending side — but still could not keep the Yankees from cutting off their eastern supply lines.

However, in spite of the heavy losses, the Rebels had withstood a major assault to hold the city and their morale was high. As both armies entrenched, a long siege seemed at hand.

By the end of August, though, Sam's unit had been moved down south of Atlanta to the nearby town of Jonesboro, where they had "a hard fight" with the Yankees. General Sherman had moved the bulk of his army to the south of Atlanta, threatening the remaining rail lines which supplied the city.

Having caught on to Sherman's gambit a day too late, the Confederates were forced to abandon Atlanta when they could not hold their last supply lines. The Army of Tennessee was not destroyed on the battlefield, but instead left town intact, its position once again made untenable by the shrewd maneuvering of a general wielding a superior force. The Bluecoats raised the Stars and Stripes over Atlanta on September 2, 1864.

When a 10-day armistice began on September 12, the citizens of Atlanta began streaming southward. On September 19, Confederate President Jefferson Davis slipped in to visit his defeated troops in the town of Palmetto, only 20 miles southeast of Atlanta and virtually in the shadow of Sherman's victorious army.

Inspired by Davis' visit, the Confederates moved around west of Atlanta and began a northward march in an attempt to threaten Sherman's supply lines from Chattanooga and draw his army away from the city. At first, Sherman did send most of his troops after the Tennessee Army.

As the Rebels moved northward towards Chattanooga, they easily recaptured several lightly defended points along the Chattanooga-Atlanta railroad, capturing 600 Yankees and 200 negroes by the time they reached the town of Dalton in northwest Georgia. However, with a sizeable enemy force closing behind and Union-controlled Chattanooga ahead, the Tennessee Army could not hold the railroad. Instead, they turned westward into Alabama in mid-October.

Sherman had in the meantime secured approval for his infamous "March to the Sea". He returned to Atlanta with part of his force to make preparations for the rampage to Savannah, leaving 60,000 men to pursue the Tennessee Army, which by this time had been reduced by over a third and numbered only 40,000.

By November 30, 1864, the Army of Tennessee was positioned for its last major engagements of the war at Franklin and Nashville, Tennessee. In the Battle of Franklin, 22,000 Rebs charged entrenched Union positions. Although they succeeded in forcing the Yankees back 18 miles to Nashville, the cost was terrible. Twelve generals were killed as the Confederates suffered 7,000 casualties — three times those of the enemy. At the Battle of Nashville two weeks later, the Rebels were decisively defeated when Union soldiers with new rapid-firing carbines forced the tattered army into wholesale retreat.

As thousands of Rebels surrendered, thousands more dropped their weapons and supplies to make a faster getaway. Union calvary men gathered their horses and for two weeks pursued the retreating remnant army from one river to the next across the Tennessee countryside.

Half of the Confederates were without shoes in the December cold. Deserters and stragglers by the hundreds were left behind. Only the rearguard fighting of Confederate General Nathan Bedford Forest bought enough time for the exhausted army to escape across the Tennessee River and into Mississippi.

By the time the shattered army reached Tupelo, Mississippi on January 10, 1865, only 20,000 ragged men remained from the fighting force of 65,000 which had first faced General Sherman in the north Georgia Hills below Chattanooga.

From Tupelo, what was left of the Army of Tennessee set off on a thousand-mile trek through the heart of Dixie to join with other Rebel forces in the east for a desperate last stand against the surging Northern armies.

Over the next two months, Sam recorded a series of long marches, an occasional train ride and one river-boat ride as the Tennessee Army traversed Mississippi, Alabama, Georgia and the Carolinas to draw close to General

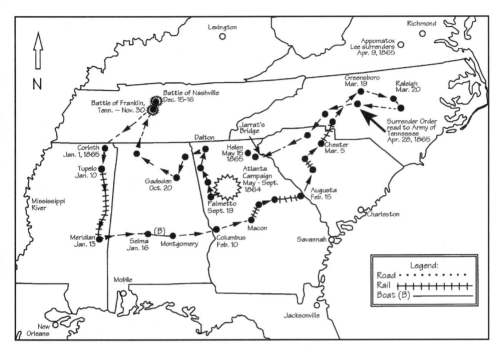

THE CIVIL WAR JOURNEY OF SAM CONLEY
May 30, 1864 - May 15 1865

Robert E. Lee's Army of Northern Virginia by mid-March.

However, Union forces kept the armies apart. Ironically, General Sherman's forces were not far from the Army of Tennessee. But, sensing the end was near, Sherman chose not to engage his old foe. On April 7th, Sam's diary records that the Tennessee Army learned Richmond was being evacuated. General Lee surrendered for the Army of Northern Virginia at Appomattox on April 9, 1865.

On April 28, an order was read to Sam's unit saying his Army had surrendered. The soldiers of each state would be marched home by their officers. The next day, the Rebel soldiers turned in their arms.

On May 3, Sam started on march for Georgia. His unit received their paroles in Greensboro, NC on May 5. By May 13, he had reached Jarratt's Bridge and the start of the old Unicoi Road near Toccoa at the Georgia-South Carolina border. Two days later, on May 15, 1865, Sam Conley took the last few steps on his long and dangerous journey, reaching home three weeks shy of his 21st birthday.

Sam was soon busy finishing his schooling and working the fields on the Conley place. His notes had chronicled the last campaigns of the Army of Tennessee, but made no complaint about the hardships and defeats suffered along the way.

Sometimes, though, travails come in bunches and Sam's were not quite over. His diary notes that he "took sick with smallpox Dec. 16, 1865", a week after returning from a trip to Atlanta. A month later, his little-sister Hattie died. Whether or not her illness was related to Sam's the diaries do not say. However, they do show that Sam missed her. Eighteen years later Sam still thought of Hattie, for in his diary of 1883 he wrote a simple notation: "In memory of Hattie J. Conley, born March 2nd, 1847, died January 13, 1866, . . ."

FARM LIFE

Though hardly a white-columned plantation, the Conley place was somewhat larger than the typical mountain farm. In 1828, Henry Highland Conley paid a thousand dollars for 125 acres of land along the east side of Chattahoochee River. This tract included all of the area where old downtown Helen was located and what today is known as the Unicoi Hill and Pete's Park in the middle of town. To the flatter portions of this sizeable area the Conleys applied not only their own labor but also that of the slaves who toiled there for forty years.

After the war, the blacks living on the Conley farm and the England place across the river left the Helen valley. Although they may have remained for a short time after the war, they were gone by 1867. When the former slaves moved on, hired hands were needed on the Conley place. Prominent among them were Sam's friend and fellow veteran Joel Abernathy and his family, who appear to have lived on the Conley place for a time.

Especially during the growing season, the Conley farm might have appeared to the sensitive eye of an artist as a sort of organic quilt. Sam's diary lists a dozen different sections, or "pieces", under cultivation, ranging from the "swamp piece" and the "bottom strips" near the river to the "hill piece" and the "upper cut" on the Unicoi Hill. Corn was a mainstay, but a variety of crops were grown. Many vegetables probably found a home in the "garden piece" near the house, while the peach and apple trees in the orchard at times had a carpet of oats for company.

The Conley house was a log cabin which sat atop the hill, overlooking the Unicoi Road. A split-rail fence enclosed the yard, which was probably bare earth and swept to keep it clear and smooth. The cabin was finally torn down in 1913 to make way for the Mitchell Mountain Ranch Hotel. Although no sign of the cabin remains, Conley descendants claim that the huge old hemlock fir which today stands upon the Unicoi Hill was planted by their ancestors.[4] The two slave cabins probably stood in the bottom areas closer to the river. Sam's diary also mentions a crib where corn was kept and a storehouse. There were small barns or stables and probably a smokehouse for curing and storing meats. Sam had a farm wagon and a buggy for family transportation.

Sam worked a lot on the split-rail fences used to guide the larger farm animals, which included cattle, horses, hogs and mules for plowing. Many of the cattle and horses were taken to the mountains and left there to browse for the summer, a mountaineer practice which continued well into the 1900s. The early days of such browsing in a nearby mountain county were described as follows:

> Early settlers drove their livestock, in the spring of the year, out onto the ranges. The grass and the peavine were succulent and nourishing. The stock got sleek and fat during the summer.

> "On some level place, from a fallen tree, a hollow was chipped in the top of a log. This was called a lick log. Salt was placed in the hollow for the stock. Once a week or so pioneers went to the mountains to salt their stock at the 'lick'. Here the horses and cattle gathered at nightfall as if about the familiar bars at home. This was the centre of the circle of their day-long wanderings. The horn, voice or bell of their master often signalled them in to this focal point.

> "When attacked by wild beasts, the cattle made terrific defense. The bleating of a young calf or similar alarm might in a moment rouse the whole herd. Gentle milk cows and patient oxen accustomed to the yoke, with hair raised, broke into the insane fury of a stampede. All of them, bellowing and blowing, fought as recklessly as mad bulls. They thus resisted such savage foes as wolf packs and enemies even larger and more dangerous.

> "Horses did not defend themselves so well. Their worst enemy was a panther crouched upon a tree, from which it might spring upon some unsuspecting animal passing below. (Some pioneers were warned by the falling of fine bark upon them just underneath such a creature.) Colts especially were the objects of such attacks. Many a time, upon horses that had run away and survived such seizures, I have seen scars of panther claws.[5]

MOUNTAIN FARM. Taken about 1890, this picture shows the farm of Andy Adams located a mile above the Helen valley. Both the setting and the set-up are very similar to Conley place during Sam's day. Mountain farms typically had several smaller outbuildings rather than one large barn. Two types of split rail fence are shown. In the days before lawn mowers and weed trimmers, things took on a ragged appearance during the growing season. Twenty five years after this picture was taken, a son of Andy Adams advertised in the county paper, seeking work taking cattle to the mountains for the summer.

Sam doesn't mention such attacks, probably because the early settlers had effectively targeted the large predators. By the time Sam's diaries were written, the wolf had probably been hunted and trapped to extinction and the number of panthers greatly reduced.

Like most of their neighbors, the Conleys had sheep, one of which they reported to the county tax man as having been killed by wild dogs in 1864.[6] At the time, the authorities seem to have been checking on the dog situation, since the tax man also recorded that the Conleys had two dogs of their own. Chickens and sometimes geese were also part of the scene on the Conley place.

Sam's older brother John Conley lived across the river with his wife and five daughters in a house close beside the Unicoi Road and located about where Wendys restaurant and the White County Bank stand today. The inventory of John's estate upon his death in 1890 tells something about the state of worldly things at the time, when his 125 acres of land on the west side of the river were valued at $1000 and the balance of his material possessions at $198.75.

The inventory consisted of only 36 different entries including the livestock and a set of carpenters tools, but they were enough to manage the affairs of house and farm. Along with some simple furniture and kitchen goods, the house contained four lamps for light, two looking-glasses (mirrors), and ten pillows and 30 bedspreads and quilts to keep the occupants of the four beds warm. One can imagine that it was often quiet in the house except for the crackling of the fire and the ticking of the clock, which was valued at fifty cents.

One thing missing from the list is a gun, an item which was on almost every estate inventory done in the county until the last one was taken in 1925.[7] The typical gun in Sam and John's day was a muzzle loading rifle, valued at about five dollars. Even in the 1930's, black powder and rifle balls could still be bought in the mountains. By this time, though, more modern breech loading rifles were rapidly replacing the remaining old muzzle loaders, which were eagerly bought up by tourists at the still reasonable price of about five dollars.

The Conleys had local markets for their farm goods down at C.L. William's commercial establishment in Nacoochee Valley, and later at neighbor George Slaton's store just down the way in the Helen valley. In addition to the products of the field, Sam at times recorded sales of eggs, hens, meat and hides. Sam often traded with his neighbors and also with brother John, who paid him $3.75 for fodder and shucks in May of 1883. Sam and Joel Abernathy also had cooperative arrangements whereby Joel traded his labor for a share of the bounty from the Conley fields.

The diaries indicate the Conley farm didn't generate a lot of income, not more than a few hundred dollars a year. But back when a mountain farm was largely self-sufficient, it didn't take a whole lot of money to get by. Sam's total personal expenditures for 1883 came in at $143.30, including $13.75 in train fares and fifty cents to get a tooth filled. Fillings were a new technology and dental work a painful experience, but Sam probably had fewer cavities than his modern counterparts since he had little access to refined carbohydrates, or sugar.

Sam had a thermometer and often made notes on snows, heavy rains and floods, hot and cold spells, the occasional cyclone which was near enough or bad enough to hear about, and even an earthquake which occurred when he was visiting across the state in northwest Georgia. Weather was important to a mountain farmer, and it made more of a difference for those who huddled beneath the stacks of counter-paines (bedspreads) and quilts on a cold winter's night and whose travel came to a halt when the creeks got too high. Many old people around Helen have always said they remember the creeks having more water and the winters being colder than they are now. This may just have to do with the way some recollections get larger as they grow older, but Sam's diaries do give the same impression.

While there was always something to be done on the farm, some times were busier than others. Things were slower in the winter, but Sam and his helpers worked long days and even at night when it was time to plant. There'd be somewhat of a break in the late summer months when all the planting was done, and then it was back to hard labor again during fall harvest season. Even at the busiest times, though, it appears that most things could wait a day or two if need be, for Sam always found plenty of time to visit locally or make more extensive travels if he so desired.

Sam's surviving diaries cover various years from 1864 to 1887 in the Helen valley. Using excerpts from each diary with Sam's original spelling, here is a composite view of a year on the Conley farm:

January: Shucked corn. Sorted potatoes — 7 1/2 bushels soured. Hauled fodder out of upper field. Hired Ben one month, wages $5.00. Hired Charlie two weeks, wages $2.50. Went with Ben to Cleveland after oats [9 mile ride each way]. Picked geese. Took butter to store, butter to date 14 3/4. Commenced breaking up. Joel and I commenced fencing. Fixed yard fence. Hauled up oat straw. Joel and Ben three days work on fence and rails. Two lambs came. Sheled corn.

February: Hauled out 13 loads of manure on potatoe patch and roasting ear patch. Up to date Joel and Ben worked at rails and fence 7 days. Hauling rails and boards from Abernathy place. 450 rails made to date. Went to store with butter and eggs. Commenced to sow oats - 4 bu. winter and 1 1/2 rust proof. Heifer [named] "Duck" died — three more sick. Fencing pasture. Went to A.P. Williams [store] in P.M. collected $36. Planting pepper squash & cucumbers. Worked Selum [a mule] for the first time.

March: Planted Irish potatoes. [Heifer named] Moon Shine died. Hauled manure in garden. Planted garden. Nell had a colt. Commenced to planting corn. Sowed 10 more bushels winter oats & 3 bu. rust proof. Plowed some and burnt off lower field. Burnt off left swamp. Planted english peas. Planted onion buttons and beets. Set hen 13 eggs. Sallie cow stayed under bank all night. John, Olivia, and I brought up Sallie cow. Finished fencing. Cross breaking the Abernathy strips. Cleaned off yard and cut pines along road. Belled the cattle and went to mill for cows.

April: Hauled out sand & hauled two loads manure for melon patch. Planted corn. Began to plow over second time. Planted corn and beans. Planted Abernathy strip. Finished breaking orchard second time. Took cattle to mountains. Fixed spring & Olivia sconsed. Planted beans, tomatoes, squash, cucumbers, & raddish. Set hen 15 eggs. Planted orchard. Took off two hens and chickens. Yams and spanish potatoes sold 44 bu. for $35. Went to Spencer's mill with corn.

May: Continued plowing over second time. Set out 180 cabbage. Colts taken to mountains. Put out some potato slips. Commenced plowing corn in swamp cut. Finished hoeing hill piece. First mess of peas. First mess of Irish potatoes. Took yearling to mountains and found my colts. Hoed melons. Finished rock pile piece.

June: Plowed corn. Commenced plowing swamp piece again. Martin boys helping hoe corn. Laid by corn patch. Put out potatoes last night and this evening, Ben worked at night. Plowed sweet potatoes. Sowed peas. Finished cutting wheat. Hoed piece above garden for the first time. Big rain — orchard oats flat on ground.

July: Finished working the third time. Tied up first oats. Hoed upper cut oats. Plowed hill and garden pieces the third time. Hoed upper cut oats. Over corn the second time. Persimmon tree struck by lightning. Finished laying sweet potatoes bye. Hauled in 1400 bundles oats. Finished tying up oats. First mess of roasting ears.

August: Made cow pen and plowed turnip patch. Sowed turnips. Got through with oats - made 14 bu. rust proof and 35 bu. winter oats. Took wheat to mill. Went to Clarkesville with oats - 16 bu. @ 40 cents [each]. Took load melons to Burnetts. Pulled a little fodder.

September: Hired Ben for 1 month for 4 bushels of corn @ $1.25 per bushel, $5.00. Paid for hiring 6 days at 25 cents per day. Brought pied cow and calf from mountains. Finished pulling fodder. Commenced to picking peas. Sowed oats in orchard, 3 1/2 bushels. Brought colts from mountains. Killed shoat [young pig]. Bought 8 geese. Charlie worked 13 days.

October: Sowed rye. Hired Ben for 1 month. Gathered potatoes around house, four loads. Sowed wheat above garden, three pecks. Sowed pieces next to yard, 1 bushel and 1 peck. Dug potatoes, 45 bushels yams, 55 of spanish. Sold Joel a cow and calf for 4 sheep, 3 hogs and 200 bundles of fodder.

November: Joel hauled first load cane. Harve Adams brought four head of my cattle from mountains. Put away cabbage. Joel a' commenced sirup.

Began Joel's cane. Finished sowing wheat. Threshed oats. John killed beef. Ben & I worked mountain road and hauled 4 loads wood. I went to G.W.S.'s [George Slaton's] with peas and hide. Cleaned up 7 1/2 [bu.] rust peas, 4 1/2 purple. Killed hog wt. 160 pounds. Gathered 4 loads of corn from lower cut and shucked.

December: Joel shucked corn. Hauled up seven loads of corn and finished. Finished shucking. John and I brought cattle from Tom McClures. Bought two colts of John Leonard. Bought yearling from Columbus Adams, $2.75 and one from Joel, $3.75. Hauled 5 loads of wood and one of pine. Killed hog - weighed 175 lbs. Fixing fence from crib to stable. Hauled in fodder - 325 bundles.

When outsiders began visiting the mountains in increasing numbers in the 1900s, most mountain folks still lived much as they had in Sam's day. There arose a perception that they were lazy and backwards, a view adopted by early film-makers and which has endured to some degree ever since. As for the "lazy" part, it could certainly appear so to a traveler fresh up from an ever-busy city who saw a lot of folks "a settin' on the porch" and not doing much else.

But the pattern of life was different in the mountains. Mountaineers did what they needed to do, and what made sense in terms of both the weather and their isolated rural economy. Most visitors came during the hot summer months, a time when it made sense to dodge the heat of the day in the shade of the porch. And with the crops planted, many farmers were just waiting for them to come in.

And as for the "backwards" part, it was true in the sense that the city-slicker was really taking a trip back through time. A 1920s urbanite came from a world where tall buildings overlooked streets crowded with automobiles, streetcars and a bustling citizenry energized by electric lights, movies, and radio. Life in the isolated mountains, though, was still much closer to that of the frontier. Far from the world of flappers and bath-tub gin, blacksmiths and wheelwrights still worked at their trades, and some areas were still thirty or more years away from electricity, phones, and indoor plumbing.

Even though the days of the family farm have passed, the pattern of life is still different in the mountains. This is in part a legacy of the old order, where folks worked hard when there were things to be done, and didn't worry about it otherwise. Also, the local economy still has strong seasonal elements and is not yet dominated by 9-to-5 occupations. Tourists, though, are now more likely to view the locals at work in a shop or hotel than to see them settin' on the porch.

TRAVELS

Modern voyagers can cover the 90 miles between downtown Atlanta and the city of Helen in just over an hour-and-a-half, if traffic allows. Hurtling along in gleaming automotive cocoons, they need never be exposed to the environment unless a breakdown occurs. Radio contact with the mother civilization can be maintained at all times, including two-way voice communications and Internet connections if so desired. When the antennas retract and the hatches open, the occupants are about as clean and well-rested as when they left, and probably mentally in about the same place as well due to the speed and insularity of the trip.

For Sam Conley, of course, travel was much different. On December 17, 1867, Sam got on his horse and set out for Atlanta. The 90 mile journey took three days. Traveling 30 miles a day, he spent the first night in Gainesville and the second in the town of Lawrenceville before reaching Atlanta on the third afternoon. Roads at the time were often only a step above trails, rough and narrow, and little more than muddy creases across the Georgia landscape during bad weather.

Since even local trips took a while, Sam often spent the night when visiting only a few miles away in Nacoochee Valley; guests were extended the same hospitality at the Conley place. Although the weather affected the comfort of the exposed traveler, Sam was not much deterred by the elements. He made extensive journeys in all seasons of the year, as likely to go in January as July. The weather could stop things entirely, though, when the creeks got too high. It happened to Sam in January of 1885, when he got "water bound" and had to stay the night with neighbor George Slaton, in sight of his own house but unable to cross the rushing waters of the Chattahoochee.

The longest trips Sam recorded in the years after the war were those he made to the town of Dalton and surrounding Whitfield County in the northwest corner of Georgia. Dalton/Whitfield County lay on the other side of the state, about a hundred miles distant across the rugged by-ways of north Georgia. Sam took several different routes, all of which traversed rough mountain areas. He crossed the highest ridges via commercial turnpikes, which were improved roads chartered by the State but run by private operators.

The trips to Dalton took three or four days, but, judging from his activities before a trip, at least it didn't take Sam long to pack. Here are his notes on a Dalton trip in 1882:

Dec. 15.   All night at Dr. Starr's [Dr. Starr lived in Nacoochee valley about three miles from the Conley place.

In 1847, Dr. Starr performed the world's first operation to remove a kidney stone, performing the surgery in the Dahlonega courthouse.[8]]

Dec. 16.   All night at JDLs. [Brother-in-law John D. Leonard, who also lived down in the valley]

Dec. 17.   Returned home by Presbyterian Church. Heard Mr. Smith Preach. [This church was in Nacoochee, founded by a middle-Georgia man after the Civil War].

Dec. 18.   P.M. started for Whitfield County. Traveled 6 miles and stayed all night at Lem Allison's.

Dec. 19.   Cold and cloudy. Traveled thirty miles and stayed at Mr. Parks 11 miles beyond Dahlonega. Lodging $1.00.

Dec. 20.   Cold and sleeting. Roads were blockaded with trees. Made 30 1/2 miles and stayed at Osborne's. Bill $.60. [Osborne lived in Gilmer County near the present day town of Ellijay]

Dec. 21.   Pikage $.25 [This turnpike crossed the Cohutta Mountains]. Made 29 miles and reached Mr. Gregory's 4PM [Mr. Gregory was Sam's brother-in-law, married to sister Martha. The Gregorys lived near Chatsworth. Sam reached Dalton the next day]

Although Sam usually went by horse, by the 1870s there was an alternative and much faster way to get to Atlanta and on up to Dalton and Whitfield County. In 1875, a rail line was completed from Atlanta to Charlotte, NC. At its closest point, the line passed through the newly established town of Mt. Airy about 25 miles southeast of the Helen valley. Built as a resort, Mt. Airy was so-named by its developer in tribute to its "fresh mountain air". Rising far above its previously unheralded status as a bump on the Georgia piedmont, the new town was proclaimed in true developer fashion as a notable spot for its many springs, charming views of mountain scenery, relief of hay fever, and "salubrious" atmosphere.

The grand white-columned hotel at Mt. Airy was evidently a happening place, for Sam sometimes went there to see a preacher or other entertainment, or just to enjoy the salubrious resort life for a few days. Mt. Airy was also where Sam occasionally caught the train for trips to Gainesville, Atlanta, and Whitfield County. While the train offered a fast way of getting around, it would have been faster if Sam hadn't lived most of a day's ride from the Mt. Airy Station. But even with the 25 mile ride from the Helen valley, Sam could reach Atlanta in one day instead of three, and the steam excursion to Whitfield County was an overnight trip instead of a three or four day odyssey. Train rides were expensive for a mountain farmer, though, which may explain why Sam still did most of his travelling by horse even after the train arrived.

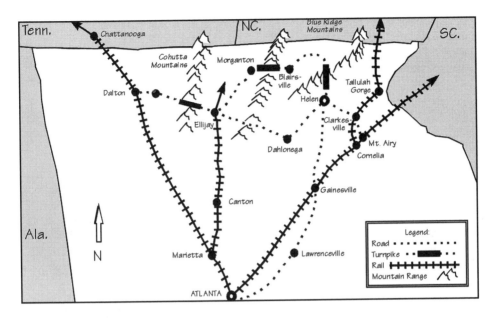

SAM'S TRAVELS. On horseback, a trip to Atlanta or Dalton took three days. Sam usually made his Dalton rides by way of Dahlonega, crossing a private turnpike west of Ellijay. On one trip, he followed a more northerly route through Morganton and Blairsville, paying tolls on three turnpikes along the way. After the Atlanta-Charlotte rail line opened, Sam had a much faster, although more expensive, alternative.

In July of 1883, Sam went to Mt. Airy to catch the evening train for a trip to Whitfield County. It's a much shorter tale than that of the same journey by horse:

July 20.	9 o'clock at night. Left for Dalton GA. Arrived at Atlanta at 1 in the morning of the 21st. Fare $2.40. Hotel $1.00.
July 21.	Left Atlanta 7:30A. Arrived Dalton 11:30A. Fare $3.00. Met Mr. Gregory there and went out and spent the night with him.

His expenses for the trip totaled $6.40, versus only $1.85 for lodging and pikage when he went by horse. Counting the time to get from the Helen valley to Mt. Airy, the entire trip took about 24 hours.

In 1882, a connecting rail line was established which ran through neighboring Habersham County and past Tallulah Gorge before eventually making its way up to Franklin, NC. At its closest point, the Tallulah Falls Railway (known locally in its later years as the "Total Failure Railway") passed through Clarkesville, only 14 miles from the Helen valley. Sam and many of his neighbors rode this train on excursions to Tallulah Gorge and Rabun Gap. On the T.F.R.R., the thirteen mile ride from Clarkesville to Tallulah Falls took 45 minutes.[9]

WAITING ON THE TRAIN. White County's long-awaited railroad did not reach the Helen valley until 1913, but for 40 years before that, trains were not too far away. Pictured here is a train on the Tallulah Falls Railway established in 1882.

The railroads brought a degree of prosperity to Habersham County as tourists arrived and new towns sprung up along the lines. This fact was not lost on White County, where the Cleveland Advertiser regularly cried out for the citizenry to rise up and obtain a rail connection of their own. In one editorial simply entitled, "WE WANT A RAILROAD", the Advertiser asked "Will White County lay still and let rail roads be built all around her?".[10] Some answered the call during the 1880s as several local groups were organized for the purpose, but none succeeded. Although White County would eventually get its railroad, it would take nearly 30 more years. In the meantime, railroads were accessible for special occasions, but most often transportation remained a more traditional affair.

Perhaps the most interesting thing about Sam's travels is how often and how extensive they were. As a mountain farmer, Sam had little need to travel very far for business purposes. Usually he was visiting, and on his longer trips went to stay with sisters or brothers who had moved on further to the west. However, the diaries also show that he was on a quest of sorts, and it was in the west that he found what he was looking for.

THE ROMANTIC ADVENTURES OF SAM CONLEY

On the afternoon of May 15, 1865, a familiar figure came into view along the Unicoi Road. The young man crossed the Chattahoochee River at the lower ford and then left the road, making his way to the cabin at the top of what is now known as the Unicoi Hill.

Sam Conley, lately of the Army of Tennessee, was home at last. He had walked three hundred miles since the surrender in North Carolina two weeks before, and had traveled over 1000 miles during the year before that in service to the Southern cause. But now that the war was over, Sam's time was again his own to manage and it was time to get on with more normal things.

For young people, two of life's biggest challenges are choosing a means of making a living and finding a mate. For Sam, the first came pretty easy. He'd seen a lot during the war, but decided to stay put on the Conley place and become a farmer. His decision may well have been based at least partly on the fact that his parents were getting older — Sam's father Henry was 66 when Sam returned home at the end of the war — and he was the only male child available to help work the farm. And in fact, this is the arrangement described in the elder Conley's will written in 1877:

> I will and bequeath to my son Samuel O. Conley all my interest in Lots of Land. . . on the west side of the Chattahoochee River containing 125 acres more or less. . . he being bound to pay all my lawful debts and take care of myself and his mother during our natural life time as his part and in consideration of the above [including] all stock cattle houses and household furniture.[11]

The second challenge took a bit longer to resolve. While some are lucky enough to find their mates early and close by, things didn't work that way for Sam. But even though he lived the life of a single man for nearly two decades after the war, it doesn't seem due to any lack of interest or attractiveness on his part, or to a shortage of available women, either. The first mention of ladies is in his diary for 1867, which contains a list of local girls apparently written in connection with church activities. Among the names recorded but not common today are those of Georgia and Tennessee Westmoreland, Manerva Allison, and Miss America Fain. Sam comes across as an active and sociable man all through the diaries, where the names of single friends and local girls are sprinkled across the pages.

While married people can settle down to concentrate on family affairs, single people have to stay out and about if they are to pursue the business of finding a partner. Locally, Sam was most often found down in Nacoochee Valley where he might attend church, visit a resort hotel like the Wyly House or see some of his many relatives there. He seems to have been particularly fond of brother-in-law John D. Leonard, at whose home he often spent the night. Sam had somewhat closer relations with several young ladies like Miss Mamie Wallace and particularly Miss Mamie Glen, but neither of these were to be the one for Sam.

As in his notes about the war, bachelor Sam provides details on times and places but rarely gives any direct insight into his inner feelings. However, relations with the opposite sex were vexing enough to inspire one of the more expressive sections in the diaries. It might be called, "Waiting for the Girls", as 39-year-old Sam Conley recorded the following in August of 1883:

Aug 12.　　　P.M. Sitting in the upper balcony of the Wyly House [located in Nacoochee Valley] - the mountain scenery is fine. The old Yonah and Tray Mountains are in full view. This is a delightful summer resort. I'm getting tired of waiting for the girls. Oscar is below and won't come up. I feel rather sold out. I'll stay ten minutes longer and if they don't put in their appearance I'll come down — Ho- Ho- here they come. 4:30 PM took a walk over to the Swiss Houses, viz. Oscar Knox, Miss Ida Wyly, Walter Wyly, Miss Laura McCalcan, My Self and Miss Minnie Wyly. Had quite a pleasant time. I remained at the Hotel until 10 O'Clock at night.

Sam's father died in 1878 and his mother three years later in 1881. Whether or not these events were the catalyst which finally put Sam Conley on the path to matrimony is hard to say. However, a year after his mother passed on, a new woman appeared in the diaries. Sam went to Whitfield County in December of 1882. On the 23rd, he and a friend called on the

Fincher girls, Miss Olivia and Miss Maggie. The visit was evidently reward-ing, for Sam returned to the Fincher's on Christmas Day, spending the night and staying for a party the next day. The entry for the day after that shows who Sam was interested in: "Attended a party at Widow Ann's. Went home with Olivia". And on December 30, after only a day's absence, he was back again to call on Miss Olivia, staying "to half past 11". Compared to Sam, Olivia was a young thing, just 20 years old while Sam was only a few months from his 40th birthday.

Sam headed home on New Year's Day 1883. Although he and Miss Olivia had obviously enjoyed quite a Christmas holiday together, five months elapsed before she again appeared in the diary. In the meantime, Sam visited Mt. Airy, attended parties, went with a friend to call on the Misses Glen, and generally seems to have slowed down only long enough to have a few teeth extracted on April 18. But Miss Olivia was certainly back on his mind by the first of May, when he wrote "A letter to O". The abbreviation was significant since Sam used shorthand only for special people like his brother-in-law John Leonard ("JDL") and Miss Mamie Glen ("Miss M"). But it was Miss Olivia who got the most special treatment, simply becoming "O" for the most part during their courtship.

An answer to Sam's letter was not long in coming, for he recorded a "Letter from O" two weeks later. After letters flew back and forth over the next two months at about two-week intervals, Sam went to the Mt. Airy train station on the 20th of July, arriving at Dalton the next day for what turned out to be sort of a curious visit. When he reached Mr. Fincher's on the 23rd, Miss Olivia was not there, the diary saying only "Mella and I called on Miss Maggie Fincher". Sam doesn't say if O's absence was a surprise, but if he did-n't know her whereabouts before he came, he learned them before he left. Olivia was apparently visiting with relatives — another "Mr. Fincher" — near the town of Canton some 45 miles to the southeast.

After buying a tie for 35 cents and visiting in the Whitfield area for the next several days, Sam returned to Dalton where he caught a southbound train. Although he was leaving town, he was still intent on seeing Olivia, even if it might be more difficult than he had expected. He went only about 60 miles south to Marietta before turning around:

July 27.	Took the train at 12 o'clock to Marietta. Arrived at Marietta 3:30. Hotel $1.50.
July 28.	A.M. Left Marietta for Canton at 9 A.M. Arrived at Canton 10:30. Fare to Canton $3.15. [Canton lies 25 miles north of Marietta and was reached via the Marietta and North Georgia Railroad]. Left there in

buggy for Mr. Fincher's 12 o'clock. Arrived there at 2:30 P.M.

July 29. Miss O & I took a buggy ride of 10 miles. Lemonade 10 cents. Left there at 2 A.M. for Canton. Horse and buggy $2.50. Took the train at Canton at 4:30 P.M. Fare to Marietta 75 cents. Arrived at Marietta at 6 P.M. Left there at 8 P.M. and arrived at Atlanta at 9 P.M. Lodging Atlanta 50 cents. Orange 10 cents. Remained there until 7:30 on the morning of the 30th. Got to Mt. Airy at 9 A.M.

It paints a pretty picture, Sam and Olivia in the buggy, him wearing the new tie and each drinking a five-cent glass of lemonade. In spite of the romance of the interlude, however, no commitments seem to have been made. The "Waiting For The Girls" episode at the Wyly House took place two weeks later, and except for an exchange of letters shortly after Sam's return, Olivia again disappeared from the diaries for almost a year.

In the meantime, Sam continued on his usual rounds of parties, visits, and trips to Mt. Airy. "Miss M" also appeared in the diary several times. Sam stayed home for the 1883 Christmas season. Although they may have lacked the spark Olivia provided the year before, the holidays were still a festive time:

Dec 19. Settled up to be in readiness for leaving.
Dec 20. Went to Gainesville. Fare 1.50
Dec 24. Returned to Mt. Airy Fare 1.60
   "   Party at Mrs. Church[s] at night.
Dec 25. Went home.
Dec 26. Party at the Hotel
Dec 27. Went to JDL's & returned.
Dec 28. Returned to [Nacoochee] valley and attended party at Dr. Starr's.
Dec 30. Went to JDL's and returned the 31st.
Dec 31. Party at Mrs. Robertson's.

It turns out that the respite from Olivia was only temporary, though, as Sam again headed west on July 28, 1883, almost exactly a year after the buggy and lemonade affair in Canton. He was accompanied on this trip by Miss Annie Glen, a sister of Miss M. Travelling this time by horse, the trip stretched across four days before they reached Mr. Fincher's and spent the night there on August 2. Sam visited around Dalton for the next week, splitting his time between the Finchers and other friends and relatives in the area.

Many of Sam's favorite people appear on these pages of the diary, including "O" (Miss Olivia), "JDL" (brother-in-law John Leonard), and "Miss M" (Miss Mamie Glen). The pages also describe a late cold spell, several trips to Nacoochee Valley, and longer journeys to Clarkesville and Mt. Airy.

Sam's diary entry for the last night in Dalton says, "Miss Annie and I went to Mr. Fincher's in P.M. Walter G. took her [Annie] home. I [stayed] all night at Mr. F's." Departing Dalton on August 9th, the ride home again took four days. Sam and Miss Annie reached Mr. Glen's place in Nacoochee Valley at 3:30 on the afternoon of August 13.

If this trip didn't cement relations between Sam and Miss O, it happened soon thereafter. A week after arriving home, Sam "Mailed a letter to O". A month later, on September 24, a "Letter from O" arrived, whereupon Sam quickly wrote and mailed a letter to "J.W.F", otherwise known as "Mr. Fincher". Somewhere between the trip and the exchange of letters there was apparently a proposal made, accepted, and approved, for over the next several months Sam was obviously making preparations as he purchased the following:

neck ties	.25	coat and vest	18.50
pr. pants	5.75	1 hat	2.75
2 shirts	3.75	1 looking glass	.40
2 prs. drawers	1.00	2 collars	.25
3 handkerchiefs	.90	2 prs. socks	.80
1 pr. shoes	6.50	1 pr. rubbers	.75
cologne	.65	1 comb and brush	1.15
1 buggy	67.00	1 buggy whip	1.50

On December 15, 1884, Sam got in his new buggy and set off to get Miss Olivia. He travelled his usual route through Dahlonega and Ellijay, making about the same time as on horseback. He arrived in Dalton on the 18th and married Olivia on the same day, the ceremony taking place at Mr. Fincher's. How stressful the event may have been for Sam we don't know, but Sam's diary entry on his wedding day shows that both he and his horse got a lot more than cold feet, and it is the only place in the diaries where he complains about the weather.

Far from having a fancy honeymoon, Sam and Olivia had little chance for time alone, but they did enjoy another festive holiday season together much like the one when they first met two years before. In Sam's usual succinct fashion, he says little about the ceremony, but is quite informative about natural events and Sam's itinerary. Beginning with Sam's departure from the Helen valley, here is the diary account of the matrimonial event:

Dec 15.	Left for Dalton. 1st night at Parks 11 miles from Dahlonega. Lodging $1.00. 2nd night at Sam Osborne's. Bill at Osborne's 60 cents. Pikage 25 cents. 3rd night at Mr. Gregory's.
Dec 18.	Cold, snowing, & freezing. Sister Mattie and I went to wedding. Suffered from cold. Ice on Mitchell's Bridge. Horse fell and had to take it out and take buggy across by hand.
Dec 19.	Sister Mattie, Olivia and I went and spent the night at Sister Mary's.
Dec 20.	Returned to Mr. Fincher's. Rain and sleet at night.
Dec 21.	Cold and sleety.
Dec 22.	Raining some.
Dec 23.	Took dinner at Mrs. Giddings.
Dec 24.	In Dalton at the Christmas tree. All night at Sister Mary's.
Dec 26.	Took supper at Mr. William Richardson's

Sam might have married late, but he was still married. He had come a long way and endured the extreme cold of his wedding day to claim his young bride. And with all of his romance related journeys to Nacoochee Valley, Mt. Airy and Whitfield County, Sam probably went considerably further in finding a woman than he did as a soldier in the Confederate army. But at least Sam's amorous quest was no lost cause, and even though they can be troublesome at times, romantic adventures are much to be preferred over the military kind.

As for "Miss M", Sam's sometime companion from Nacoochee, she never married. Miss Mamie Glen moved off to Atlanta, where she worked for many years in the Carnegie Library. Returning to the family place in Nacoochee Valley in her later years, Miss Mamie lived well into her 90s.[12]

COURTIN' COUPLE. This young couple from neighboring Franklin County is courting in a buggy similar to the one used by Sam and Olivia. The ride may have been rough, the roadside unkempt, the horse disinterested, and the weather a factor, but they did little to deter the pursuit of romantic adventures.

MOVING IN AND MOVING ON

On December 27, Sam and Olivia (after the wedding, the former Miss "O" became a more wife-like "Olivia" in the diaries) loaded their things in the buggy and headed home. They stopped first to visit with Sister Martha and the rest of the Gregorys in Murray County, staying for several days. From there they travelled a more northerly route than usual, going through Morganton and Blairsville on the back side of the Blue Ridge in order to visit a Captain Reed in Union County and Sam's brother Rufus near Hiawassee.

On this northern route, it was necessary to cross two additional turnpikes, the last being the Unicoi Turnpike which crossed the Blue Ridge at Unicoi Gap about 9 miles above the Conley place. The trip home went as follows:

Dec 27.	Went to Mr. Gregory's.
Dec 30.	Started home. First night at Mr. Osburns. Bill at Osburns 60 cents. Warm and pleasant.
Dec 31.	At Morganton GA. Bill at Morganton $2.00.
Jan 1&2.	Pikage 25 cents [This turnpike was between Morganton and Blairsville]. At Capt. Reed's.
Jan 3.	Arrived at Bro. Rufus - snowed some.
Jan 5.	Left for home. Rained all day. Arrived at home 3 o'clock. Pikage Unicoi 25 cents.

When Olivia finally stepped out of the buggy on that dreary winter afternoon, she may have been cold and wet, but three years after Sam's mother passed away there was again a woman on the Conley place. Far from home and in strange country, family lore has it that young Olivia was homesick and it's hard to imagine otherwise. Even so, she was soon making her mark on her new surroundings.

A week after the new couple arrived, there were "hard rains", whereupon the "furniture [was] damaged some". It can be surmised that, after three years of bachelor living, some things were a little rough around the Conley place. Also, the house was over 50 years old by this time. And a leaky roof seems to have been only the first thing that needed attention, as Sam was soon out purchasing a number of items which clearly appear to have been inspired by the new Mrs. Conley:

2 sets plates	1.50	1 sieve	.15
2 sets teas	1.50	1 oil can	.25
2 sets knives and forks	.50	1 lantern	.90
1 set tea spoons	.50	2 sad irons	1.00
1 set table spoons	.50	set and irons	1.50
1 sugar bowl	.40	" " "	.50
1 butter dish	.25	shovel&tongs	.50
cream pitcher	.20	2 lamps	1.00
milk pitcher	.40	2 looking glasses	1.10
salt cellars	.15	1 bolt Sea Island	4.00
sirup pitcher	.20	1 bolt sheeting	3.50
1 pepper&1 vinegar cont.	.25	2 brooms	.50
3 small bowls	.50	2 pad locks	.35
4 steak dishes	1.25	hasp & staples	.15
4 soup plates	.50	back band	.20
1 gal cup	.10	3 combs & 1 brush	1.45
strainer	.25	4 buckets	1.00
1/2 gal measure	.15	2 dinner tables	6.00
1 dipper	.50	2 small tables	4.00
1 dish pan	.40	dining room safe	5.00
1 room suit	35.00	3 chairs	1.80
1 rocker	1.75	6 chairs	4.00
2 wash stands	3.00	2 pr. blankets	6.00
5 1/2 yds. table linen	2.75	1 table cover	.75
1 quilt comfort	2.00	1 clock	4.00
1 doz napkins	.75	1 cook stove	18.00
1 churn	1.15	1 coffee mill	.75
1 cedar bucket	.60	2 poplar beds	8.00
3 mattresses	7.50	1 poplar bed	2.50
4 shades	4.00		
3 tin cups	.25		

Although these expenditures were extreme by Sam's modest standards, he seems to have handled them without difficulty. In February, he still had enough cash on hand to refinance a $200 note he had taken out with neighbor George Slaton two years before, paying half back and getting Mr. Slaton to carry the $100 balance for two more years at 8% interest.

1885 was a busy year for the Conleys. In addition to any assignments he got from Olivia, Sam had his usual work to do in the pastures and fields. Most of 1885 was a little different for him, though, in that he stayed close to home. Sam and Olivia often socialized down in Nacoochee Valley, but didn't go much further. The main reason for that became apparent in a simple

November diary entry:

> Nov 26.   Clear and pleasant after the cold morning. Baby born at 4 o'clock.

It was a girl. The diary indicates the birth was attended by a mid-wife, who was paid 50 cents for her work.

Babies are the future, but their names tend to say something about the past. This was the case with the new Conley child, for she was named "Hattie" after Sam's little sister who had died two decades before. Things seem to have gone well for mother and child, for in a couple of weeks Sam was back on his horse, travelling the Unicoi Turnpike across the mountain to spend a few nights with Captain Reed and Brother Rufus.

The winter of 1886 was cold and wet, with the thermometer reaching 5 below in mid-January. The baby was doing fine, though. By the end of February, Sam recorded three-month-old Hattie's weight at 15 lbs. Before the winter was over the Conleys were again out visiting in Nacoochee. And on August 18, the Conley family set out in the buggy for Dalton, averaging 32 miles a day and arriving at Mr. Fincher's on the afternoon of the third day. It was an extended visit, perhaps because Olivia was homesick. They experienced an unusual natural event on this trip, as Sam's note for August 31 reads: "Earth quake nine o'clock at night."

A few months later, in the fall of the year, Sam's sister Mattie and Olivia's sister Maggie came from Dalton to visit. They were there on November 21st when Hattie was baptized. The women left two days later when Sam took them to the Clarkesville Station to catch the first of the three trains they would ride on the return.

CLARKESVILLE TRAIN STATION. Sam first had access to a railroad at the Mt. Airy Station, a stop on the line between Atlanta and Charlotte built in the mid 1870's. The Mt. Airy Station was about 25 miles from the Helen valley. When the Tallulah Falls Railway was established in 1882, the Clarkesville Station was much more convenient.

1887 started out about like the two previous years, but ended very differently. Other than a visit by Mr. Fincher in May, the diaries record nothing other than the usual farm and social activities. Then, in August, the following note appears:

Aug 15.   Sold out to Col. Jaques.

The White County deed records show that Sam did in fact sell his lands — 125 acres on the west side of the Chattahoochee River — to Mrs. Maria A. Jacquess of London, England for the sum of $2500. As events would demonstrate, this was a very good price for Sam. The story told by Sam's descendants is that he and Olivia left the Helen valley because she wanted to be back in Dalton with her family. And it appears so, since this is exactly where they went. However, it may well be that the timing had to do with the generous offer from Mrs. Jacquess, it being one Sam could not refuse and the catalyst which prompted the move to Dalton.

Earlier in the year, Mrs. Jacquess had purchased the adjoining Trammell Place, a 600 acre tract, for the sum of $7000. Somewhat ironically, the man who sold her the Trammell place, L.N. Trammell, was a once neighbor of Sam's and would soon be so again, for Mr. Trammell was living near Dalton in Whitfield County at the time of the sale. Mrs. Jacquess completed her purchases by acquiring an isolated 490 acre tract on the headwaters of the Chattahoochee for $1000.

Perhaps Mrs. Jacquess was accustomed to such high prices in England, but most likely it was a latter day case of gold fever which inspired her to pay these huge sums. Whatever the motivation, her venture did not "pan out". Only three years later, Mrs. Jacquess herself sold out, when her entire holdings were acquired by George Slaton for $300, or less than 3 cents on the dollar. Although this price was about as low as her purchase prices were high, it does highlight the good deal which Sam got.

Parting with the homeplace was probably hard for Sam, but August of 1887 was a trying time in another way as well. His beloved brother-in-law John Leonard had taken sick in May. By August things were worse:

Aug 7.   Hard rains. Wales Leonard came for me.
Aug 8.   Steady rain in fore noon. I went up to J.D.L.'s in
                P.M. and staid to next evening.
Aug 9,10   No rain. Bro. John and Sister Addie went up to John
  &11.   Leonards.
Aug 12.   Col. Jaquess came. John came back from J.D.L's.
                Olivia & I went up at dark. John L. died at 4
                o'clock. Goods for J.D.L.s burial $9.20.

THE CONLEYS IN DALTON. About a year after first-born Hattie arrived, Sam left the Helen valley. He and Olivia moved to northwest Georgia where they lived for 15 years near her family and the town of Dalton. There they had a fancier house, a well-swept yard, and five more children.

Aside from pleasing Olivia, other things may have influenced Sam's decision to leave the Helen valley. People often make major changes after they are married. At 43, Sam was a little older and perhaps open to something new. With his parents gone and the passing of contemporaries like John Leonard, events may have loosened the strings which kept him at home. Looking towards Dalton, he had connections with relatives and others from the Helen/Nacoochee area who had moved there. With his many travels, Sam had always shown an adventurous streak. And, in the larger scheme of things, he was just doing what Americans had been doing for generations: picking up and moving on in the great American migrations to the west.

Although the Conleys have many descendants in the Helen area, they all derive from the female side, so Sam effectively took the Conley name with him when he left. His brother John was the last male Conley in the area, but had no sons to carry the name. However, John and his family were the last in the Helen valley to bear the name of one of the original families which settled there.

Even though Sam might have been somewhere towards the back of the westward pack when he set out on his personal migration to northwest

Georgia in 1887, he eventually closed the frontier gap considerably. Sam and Olivia lived in Dalton for 15 years. There they had five more children, two girls and three boys. But Sam was not at home when the last one was born. Ever the traveller, he was far to the west, scouting new territory.

Sam's diary entry for November 22, 1899 reads, "O.L. McHan left for Texas". O.L. was apparently just one of several locals who pulled up stakes to head that way. As had been the case in the Helen/Nacoochee area many years before, the lure of the frontier still pulled people westward. In 1901, leaving his family in Dalton, Sam rode the train to Texas. He set down near McGregor, a small town located about 65 miles southwest of Dallas.

The three oldest boys (ages 9, 11, and 13) made the three-day train ride a few months later, accompanied only by a good helping of Mrs. Olivia's chicken and biscuits. Sam was not at the station when the boys reached McGregor. Upon asking directions, they were advised to follow the train tracks out of town. They heard their father before they saw him: 57 year-old Sam Conley was hollering at his mule as he plowed a dusty Texas field. Olivia and the girls followed the next year.

Sam never owned property in McGregor, where he was still a farmer but rented someone else's land. Some of his Texas relations referred to him as the "Old Goldminer", indicating he may have tried some of his Georgia mining knowledge on the flat, dry Texas landscape. Or it may just be that, like miners of all ages, he was simply fond of recounting some of the amazing tales he brought from the Georgia gold region.

As his descendants recount from his later years, Sam required that a freshly starched shirt be standing ready before he went to bed each night. At mealtime, if anything he deemed necessary for a proper setting was missing from the table, he would refuse to say grace and simply sit in silence until Olivia identified and retrieved the offending item. Such demands may seem a bit stiff by modern standards, but they were not unusual for the men of Sam's day.

Around 1913, Sam and Olivia followed some of their children to Dallas, at the time a booming metropolis of nearly 150,000 people. Sam died there in the spring of 1929 at the ripe old age of 85. Olivia remained in Dallas until her passing in 1945 at age 83. In a way, considering that Sam was from the lush and isolated North Georgia mountains, he seems out of place in a hot Texas city. But like his earlier moves to the west, Sam's move to the city was consistent with the latest American migration: that of country folks to the city.

And even if Sam had never left the Helen valley, he would still have wound up living, if not in the city, at least in a small town. At about the same time Sam was moving to Dallas, the long awaited railroad finally steamed across White County and on into the valley where the Conleys had once lived. The new municipality of Helen was incorporated to become the site of a great sawmill. A crank telephone was soon ringing in the new drugstore. Wires strung through the valley carried electricity from the mill's steam-powered generators to new homes and businesses.

John Conley's former home was demolished to make way for the train. The old pioneer log cabin atop the Unicoi Hill which had once been home to Henry and then Sam Conley was torn down to make way for a fancy new hotel. Compared to what had been before, it was drastic change to be sure. But if Sam had still lived there, it's not too hard to guess where he might have been. He probably would have been at the train station, ready for whatever adventures lay down the tracks.

Sam and Olivia upon their retirement to Dallas.

* * * * *

MT. AIRY REVISITED

June, 1990s. About an hour north of Atlanta on the four lane, the exit sign announces "Cornelia" and "Mt. Airy." At Cornelia, the four lane becomes US 441 as it continues north. Like its predecessors, this modern highway has by-passed Mt. Airy, leaving it a bit hard to find. Older, more intimate roads wander into the old town, paved now but still showing the bends and crooks which contemporary highway engineers strive to eliminate.

The Mt. Airy Friday afternoon rush hour is peaceful, disturbed only by an occasional solitary vehicle ambling down main street. Mt. Airy City Hall stands upon a high grassy hill in the middle of town, the highest point around. In front of the building stands a huge old tree. Under the tree, a man sits easily atop a picnic table, his legs crossed and his feet resting on one of its benches. He is Virgil Loudermilk, who grins and says this is his office. Having always lived in Mt. Airy, he was policeman for 24 years and now is city manager. In the perfectly comfortable embrace of an early summer's evening, it's an ideal spot in which to work, if that is what Virgil is actually doing.

A sign in front of City Hall states that this is the highest point on the railroad between Charlotte and Atlanta. It also still proclaims the "fresh mountain air" and "fine mountain views" which Mt. Airy has been proclaiming since Sam Conley first set eyes on the fancy resort hotel there 120 years ago. As far as can be determined, it has never been demonstrated that the atmosphere thereabouts is much better than anywhere else in northeast Georgia, but neither have such details ever been shown to stop a good proclamation. It can be noted also that Mt. Airy has not come up with much else to proclaim in the years since the old hotel was built.

This hill was the site of the old resort hotels. The original one was built in the 1870s when the railroad came through. From the picture Virgil shows, it was a grand place, multi-storied with towering white columns in the middle, and certainly a bright spot for the locals. It burned about 1913, Virgil says, and was rebuilt only to burn again about 20 years later. Business was good until the Hoover Days, but then Mt. Airy declined precipitously and hasn't been a resort since. Even though its fresh mountain air and fine mountain views are pretty much unchanged, times are not.

According to Virgil, the huge old tree which shades his outdoor office is a Chinese ash, the only one remaining of the five planted out front when the first hotel was new. If Sam Conley could somehow start at this tree and ride the 25 miles from Mt. Airy back to the Helen valley today, aside from the enduring mountain views, he would find few things unchanged. His horse's feet would never touch the earth since the entire route is covered with asphalt, and the development now seen along the way would simply be inconceivable for a mountaineer of his era.

When he reached the hill in Helen where his house once stood, he would find no sign of it, nor of the Mountain Ranch Hotel which followed it. He might still think the hill a pretty place, though, for he would find it mostly in grass interspersed with a few trees. And after a while, he might realize that some of these trees were somehow familiar, not by their size and shape, but by their kind and location, for at least the leaning old hemlock fir and perhaps one or two more were there in Sam's day, planted by the pioneer Conleys many years ago and enduring yet, and like the old Chinese ash on the Mt. Airy hill, living reminders of days gone by but perhaps not really that long ago.

CHANGING TIMES. At about the time Sam Conley moved to Dallas, the railroad finally steamed into White County and on up to the Helen valley. Taken from atop the Unicoi Hill, this picture shows the old fields of Henry and Sam Conley in the foreground while steam from the great sawmill rises over the former homeplace of John Conley. Portions of the original Conley fields near the river were prospected during the Gold Rush.

MOTHER NATURE AND THE EASTERN END OF
THE GEORGIA GOLD RUSH

For something so scarce, gold really gets around: its atoms are detectable even in seawater. And from the miner's point of view, such dispersion is exactly the problem, since there are few locations where concentrations are high enough to support viable recovery operations. One favorable area is a narrow band of gold which runs from southwest to northeast across the entire state of Georgia. Entering from the Alabama piedmont in the vicinity of Carrollton, it continues in a fairly straight line to Rabun County in Georgia's very northeastern corner, where it departs for Carolina.

Although the precious yellow metal has been found in spots all along its length, the richest part by far has been the section known as the "Dahlonega Gold Belt", an area about thirty miles long and up to four or five miles in width. The gold rush town of Dahlonega lies on this belt's western end, its name taken from the Cherokee words for gold. While Dahlonega receives most of the attention, the Helen valley and the neighboring vales of Sautee and Nacoochee contain the eastern end of this rich belt, and have been the scene of as much prospecting and speculating as any other part of the north Georgia gold region.

There was an early difference, though, between the two areas. The Dahlonega Gold Belt was once entirely in Cherokee possession, but by the Treaty of 1819 the Indians unknowingly relinquished the eastern half. Thus, when the Gold Rush began a decade later and gold seekers at the Dahlonega end of the belt were charging onto Indian territory in free-for-all fashion, those at its eastern end were forced to deal with white land owners in the Georgia county of Habersham. And in contrast to the melee west of the Chestatee River, the antics of would-be miners at the Habersham end were somewhat further constrained by the civilizing institutions of church, government and community which the pioneer settlers had already labored for nearly ten years to build. In the Helen valley, the Englands, Bells, and Conleys resided on sites where gold was waiting to be found.

Although tales of Spanish mines and lost Indian treasure caves permanently reside in north Georgia lore, the discovery which initiated the Gold Rush was not made until white settlers had been in the area for nearly a decade. And almost from the start, there have been rival claims about when and where north Georgia gold was first found.[1] Nobody has ever argued about it much, though, as for years local historical markers and chambers of commerce have simply promoted their own versions without bothering to disparage the competition. Such an approach is quite in keeping with tradition in the gold region, where over a century-and-a-half of gold inspired lies have created a certain feeling that it's best to simply discount most miner's tales and then judge them on artistic rather than factual merit.

At least two claims place the initial find on Dukes Creek, a large Chattahoochee tributary which runs several miles west of the Helen valley. Other well known accounts place the original discovery at different points near Dahlonega. Some versions say the find was in 1828 or earlier; if so, it took a while for the word to spread, for Helen area deed records do not show an increase in activity until late in 1829. Georgia newspapers began reporting the discovery at about the same time.[2] In any event, a few miles and a few months notwithstanding, the great Georgia Gold Rush was in full swing by the start of 1830.

Georgia's gold fields were not the nation's first. Gold had been found during the 1700s in several Southeastern locations, and for nearly thirty years before the Georgia rush, significant quantities were being mined in the North Carolina foothills. In fact, these earlier finds were certainly a factor in the Georgia discovery as they had newcomers looking for that telling yellow glint in the newly settled lands.

After a slow start, thousands of people eventually flocked to the Carolina diggings, giving rise to boom towns and large mines where as many as thirteen different languages were said to be spoken.[3] Even so, north Georgians over the years have determined that theirs was the first "real" gold rush, largely on the basis that it was more intense and larger in scale. Local pride plays a factor, but Georgians have a case which can prevail on artistic merit if nothing else, and the Georgia experience was certainly a phenomenon.

As thousands of would-be miners poured into the area, intense competition resulted in a shortage of mining sites. According to a description supplied by a US army major, valleys on the Cherokee territory around Dahlonega filled with a mixture of "whites, Indians, half-breeds and negroes".[4] The cast of characters in Habersham County was similar except that the Indians and the half-breeds were not among them. Where it was necessary to get the permission of white land owners, desperate men in some cases paid substantial sums to rent mining areas for periods of 40 days or less[5], literally not enough time to do much more than "scratch the surface" for all but the richest and most accessible deposits. The army major further described one gold-crazed group, who:

> . . . presented a most motley appearance. . . , boys of fourteen and old men of seventy. . . , comprising diggers, sawyers, shopkeepers, pedlars, thieves, and gamblers, etc. besides them were also found in the hopeful assemblage two colonels of Georgia Militia, two candidates for the legislature and two ministers of the Gospel, all no doubt attracted thither by the love of gold.

Records from the Helen area show that few were immune from gold fever, for everyone from indigents and poor dirt farmers to doctors (including one who mortgaged his doctor bag), prominent Southern politicians, wealthy eastern businessmen, and prosperous lowland planters showed up looking for their share of the wealth. In most cases, the love of gold soon gave way to feelings of despair as time, toil and money disappeared into diggings which yielded little return. Excerpts from a series of letters sent home by one dispirited miner describe such feelings, which got so bad that beauty itself disappeared from the Georgia hills:

> I presume from your not coming that the gold fever, which you showed strong symptoms of taking when in Habersham, has entirely subsided and good luck to you that it has — for disappointment, vexation, and loss would have been the result of your coming here. . . . You can form no idea of the exaggerated accounts there is reported of this country 80 or 100 miles off. Gold they think can be had, to the satisfaction of avarice itself by a little labour. Such reports has drawn a multitude of persons to this country with the full calculation of getting rich. From the number therefore wanting employment it makes lots of value very hard to obtain. I have never before been amongst such a complete sett of lawless beings. I do really think for a man to be thought honest here, would be a disadvantage to him, or at least he would be set down a fool and treated accordingly. . . . 4/23/1833

> The mining business in this country is very dull. Everybody is getting tired of it. Those who have nothing to do with it are best off. . . . 11/6/1833

> I shall stay here but a few days longer when I hope to leave the country forever. . . . It has been raining, snowing, and hailing here for the last week. . . . when this country is in its most flourishing condition, it's not overly pleasant. . . . 1/30/1834[6]

In spite of the bad experiences of many miners, gold in north Georgia was "quite freely if not abundantly found".[7] Although years of prospecting failed to reveal dramatically rich deposits like those later discovered in California and other western states, Georgia mines generated thousands of pounds of gold, enough for the establishment of a federal mint in Dahlonega. And although there were ups and downs, significant mining activity continued in the Georgia foothills for over a century.

Although most gold has been recovered from streams and valleys, bedrock is of course the original source for the metal. Gold can be widely disseminated in some rock types, but in north Georgia rock, most if not all of the gold is concentrated in quartz veins.[8] Rock miners have dug trenches and tunnels to follow these "auriferous" (gold-bearing) quartz veins which can be thinner than a razor blade or more than 6 feet in width. Such veins can be frustrating, for the gold often teases with rich "pockets" before "pinching out" to leave the hardworking miner with no clue as to when or if it will reappear as he burrows into the mountain.

In some areas the hard underlying bedrock is covered with a transitional layer known as "saprolite", or soft, thoroughly decomposed rock which has been penetrated by water but not yet transported or converted into soil. Such layers can be extensive and over 75 feet deep. In saprolite, the quartz veins are usually still intact since they are much more resistant to decay than the surrounding rock. Rock miners have traditionally favored these areas of rotten, crumbly rock for vein mining since they were much easier to work than the hard, unweathered bedrock underneath.

Geologists believe the Georgia mountains were once many thousands of feet higher than they are today. As these great mountains wore away and released their gold from the underlying veins, much of it was concentrated in secondary deposits known to miners as "drift gold" or "placer deposits". These collected deposits can be found anywhere below their sources in the original rock.

Once freed from its source vein, gold is smoothed and rounded as it moves down the mountain. Nuggets which are not so worn may indicate the nearness of a vein. Secondary deposits on the side of the mountain may be close to their original sources or concentrations left by an ancient stream; either way, they are on their way down to eventually rest on the bedrock of the valley floors, where the largest placers are found.

Although there is a great deal of variation, the "pay streak" in valley placers averages about 12 feet below the surface, covered by eight feet of soil and another four feet of sand and gravel. According to long-time Helen miner Jim Vandiver, the drift gold in such placers is typically found only in the bot-

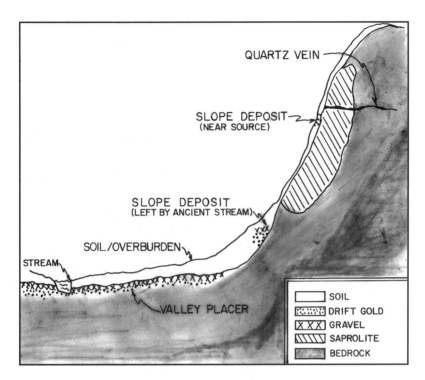

MOUNTAIN CROSS-SECTION SHOWING GOLD DEPOSITS. Early miners worked along the creeks where the necessary water was readily available and erosive action had concentrated the richest deposits of gold. On the valley floors, gold is typically found close to the bedrock under a covering of soil, sand and gravel averaging about 12 feet in thickness. Slope deposits were worked later using water from mining ditches run high on the mountainsides. Cuts and tunnels were dug to follow quartz veins through soft, decomposed saprolite and into the hard, unweathered bedrock.

tom 12 inches of sand and gravel.[9] The covering of sand and gravel is usual-ly thinnest near the creek since mountain streams run along the bedrock, and from there the overburden gets progressively thicker towards the foot of the mountain. On the mountainsides, the soil again gets thinner and deposits of saprolite are encountered (see diagram).

In the early stages of the Gold Rush, stream-side placers were the most worked deposits. Miners could easily find promising areas by working along the creeks until nuggets or "colors" (gold dust) appeared in the bottom of the mining pan, and then use the friendly waters to work the auriferous sand and gravel mixture scraped from the neighboring bedrock. Various types of sluices were usually used to process large amounts of the sandy ore mixture[10], although early miners sometimes simply ditched waterflows directly across promising areas which were then raked and checked for coarse gold when things dried out.

Using a simple technique described as "ground-sluicing", these miners are washing auriferous sand and gravel to see what they can find. Only coarse gold could be recovered in this way.

Like gold pans, sluices utilize water and rely on the heaviness of gold to separate it from lighter materials. Although there are many different designs, the basic idea is for the water to sweep away the sand and gravel as it rushes through a box or hollowed log, but to trap the heavy gold in the bottom by means of raised "riffles" or other devices. The washing techniques used by the early miners let a lot of gold get away. In fact, as recovery systems improved over the years, the same piles of material were in some cases profitably worked for a second and even a third time.

Although they were less common at first, there are many vein workings in the Helen area. Pick and shovel would usually suffice on the soft saprolite deposits, but ore was difficult to wrest from the hard bedrock with the hand tools available to early miners. Either way, the quartz ore recovered

SLUICE BOX. Shown here in a stream-side placer mine, sluice boxes take advantage of gold's high relative weight to separate it from lighter materials. While rushing water sweeps away soil and sand, the heavy gold is trapped by raised "riffles", which are typically strips of wood nailed crosswise along the bottom of the box. Since they catch both nuggets and fine gold, sluice boxes are much more effective than simple ground-washing methods.

from veins had to be crushed to free the gold it contained. Most crushing was done with stamp mills, which operate by repeatedly lifting heavy weights or "stamps" to a certain height and then suddenly releasing them to crash down upon the rock ore below. Early stamp mills were small, water-powered affairs made primarily of wood, but big mills of steel and iron with as many as 20 large stamps were used in the Helen area by the late 1800s; some of these were driven by steam.

Since mercury or "quicksilver" will bind with gold to form an "amalgam", it was used to capture the very fine particles known as "gold dust" or "flour gold". As a result, modern day panners can still find beads of mercury in some area creeks. The quicksilver would become harder as it rolled around the pan or sluice and picked up gold. Separation could be accomplished by using acid, but usually the quicksilver was boiled away, a process which produced very dangerous fumes. In later years, mercury-coated metal plates and improved chemical processes were used to further enhance recovery percentages. The following describes the use of mill pounding and mercury during the 1880s to recover gold from vein ore:

> . . . [the rock ore] is then thrown into the mills where it is pounded as fine as flour, and all the gold separated from inferior substances. A sluice of water is continually bearing the pounded earth over a copper plate upon which quick-silver had been rubbed. The mercury has an affinity for gold and takes up every dust of it, allowing the dirt and sand to pass away. On Saturday evenings, these copper plates are scraped off, the amalgam is placed in a retort and burnt off, leaving the pure, yellow gold. . .[11]

Although this account describes an interesting way to spend a Saturday evening, the plates were less than totally efficient and still allowed a significant percentage of the gold to escape.

AMALGAMATION PLATES. Rock ore had to be crushed to free the gold it contained. Crushed ore could be run through sluice boxes, but metal collecting plates developed later were more efficient. The plates were coated with mercury, which was periodically scraped off and then boiled away to separate it from the gold dust it collected.

After the initial euphoria passed and much of the "easy gold" was recovered from the stream-side valley placers, mining declined in the Georgia hills. Word of the discovery of gold in California in 1849 and later discoveries in Colorado caused a further contraction as a number of local miners headed west.[12] This state of affairs was described by pioneer miner Dr. Matthew Stephenson, a noted authority who also said that gold was first discovered on Dukes Creek in Nacoochee:

> The mines at Nacoochee, where gold was first discovered in 1828, have yielded very good profits, and in many cases they were enormous, ranging from two thousand to three thousand dollars per hand, without a dollar of capital invested. This was made on leases in the deposits in the small streams, easily and cheaply accessible; but they have long since been exhausted, and now it requires capital to prepare and open the alluvial [sedimentary] deposits of river bottoms or lands, and to work the veins which, by their decomposition, have formed the deposits. It requires engines, pumps, mills, &c., which, in the present impoverished condition of the country and people, is wholly impracticable. . . . Owing to the want of the means, all of our miners and laborers have gone to Colorado and California, where they can get better employment, and accumulate enough to commence again.[13]

As usual, Dr. Stephenson's comments showed considerable insight while demonstrating his unwavering belief in the prospects for Georgia mines. Although many miners never returned, new techniques and an influx of capital did eventually spark a mining revival in the Georgia hills.

The need for water kept early miners close to the creeks, but new approaches greatly expanded the mining fields. Beginning in the late 1850s, large ditches were constructed along the major streams of the gold belt. Low ditches were the simplest in concept, for they merely diverted a stream to a new route down its valley. By borrowing a creek and running along the foot of the mountain, these low ditches conveyed the waters needed to wash valley placers far from the stream's original channel.

In more dramatic fashion, other ditches constructed high on the mountainside supported a new process called "hydraulic mining" or simply "hydraulicking". Using water piped down from these high ditches, hydraulickers used a device known as a "Little Giant" or "water cannon" to direct a narrow, high pressure stream of water at the mountainsides and simply blow them away. With the high ditches, many previously unaccessible slope placers and even saprolite deposits could be hydraulicked very effectively.

FLUME. A flume section on the Hamby Ditch, which captured the waters of Dukes Creek. The flat boards on top made things easy for the "ditch walker", who regularly checked for leaks and debris. Mining ditches were surveyed to be as level as possible; the Hamby Ditch had a fall of five feet per mile. Its builders made every effort to avoid the use of flumes and trestles, preferring instead to dig along the side of the mountain where possible. A flume carried the Hamby Ditch across the Loudsville Road/GA Alt. 75. The Hamby Ditch was rebuilt several times; a few sections of the old flumes were still standing in the 1970s.

The hydraulic process sent tons of debris running down the mountainside to be pounded if necessary and run through the sluices. Although considerably larger and mounted on a swivel, the Little Giant was basically like the nozzle on a garden hose, except that it required a strong stand to hold it in place against the tremendous back-pressure exerted by the exiting water, and a long handle with leverage enough to aim it.

Hydraulicking was necessarily a large scale operation, both in terms of the lands involved and the financing required. To get water high on the mountains where it was needed, the ditches captured a stream far up on its headwaters and then ran for miles along the mountainside staying almost level. As they proceeded down the valley, the nearly level ditches could attain altitudes hundreds of feet above the more rapidly descending creeks. The ditches were dug into the slope if possible, but wooden flumes and even long "u" shaped metal tubes were used to cross valleys and ravines where necessary. Old timers have said that these ditches could be traveled by boat and that brook trout lived in them.

Once a high ditch was constructed, water could be withdrawn through metal pipes or, later, flexible hoses and run down the mountain to the Little Giants waiting below. Since their effectiveness was determined by the amount of pressure available, the water cannons worked best when they were at least

BUILDING WATER PRESSURE. 16" iron pipes convey the waters of the Hamby Ditch to a mining site below. Its waters were applied in many places. The ditch was breached or "gated" at numerous spots along its length for connecting pipes and side ditches as it ran for miles above the gold fields along the Hamby Mountain ridge.

fifty feet or more below a ditch. Areas closer to the high ditches were worked by simply turning the waters down the mountainside, but this was considerably less effective since even with help from the rushing water, the miners had to help free the gold-bearing soils and saprolite by hand. Water from the ditches was also used to power stamp mills.

In a number of cases, these ditches ran so high on the mountains that the waters of one creek could be diverted through a gap into another watershed. Since most ditches ran for miles, hydraulickers typically had to obtain property, leases, or easements from a number of landowners, even from those who held lands below the mines and were therefore subjected to the debris or "tailings" which washed down from such operations. Hydraulickers were often embroiled in lawsuits, sometimes to the point of being forced out of business. Considerable hydraulic mining was done in the Helen valley, where sections of two high mining ditches and a low one can still be found today. The biggest of these was the Hamby Mining Ditch, a high canal which diverted the waters of Dukes Creek into the Helen valley for years.

HYDRAULIC MINING. Connected to pipes conveying water from ditches high on the mountainside to build pressure, "Little Giants" had the power to slice through soils and even saprolite to free auriferous material. Hydraulic "cuts" were made along both sides of the Helen valley using the captured waters of Smith and Dukes Creeks.

Although the first mining ditches were constructed in the 1850s, the Civil War brought operations to a near halt. Ditch mining resumed in earnest after the war. The post-war period saw a number of large mining companies operating in the Helen area, working and re-working the valley placers, hydraulicking and washing the mountainsides with the high ditches, and in favorable spots making extensive cuts and tunnels through the saprolite and bedrock to free gold-bearing quartz. It's said locally that, since a number of these companies were financed by northerners and so many lost money, the situation was referred to as "The South's Revenge" against the Yankees.[14]

Except for the interruption caused by the war, mining activity in the Helen area was nearly continuous until about 1900. But as production costs increased while gold prices remained stable, the big companies were squeezed and eventually disappeared, mostly leaving the locals to prospect among the abandoned workings of the gold region. Activity bottomed out shortly after 1900 and remained flat until the 1930s. However, the gold mining days were not quite done. In 1933, the federal government raised the official price of gold by 70% (from about $20/ounce to $35), sparking a decade-long burst of activity in the Georgia gold region.

Several Helen area miners at this time were large sewer contractors who were motivated at least partly by the desire to employ equipment which otherwise would have stood idle during the dark days of the Great Depression. One of these was E.L. Gedney, who heard about Georgia gold while doing a job in Columbia, SC. There, a trusted fellow contractor told him that he was "making all kinds of money" with a gold mine in White County, since the price of gold was up and labor was cheap there. Taking the message to heart, E.L. moved his equipment to White County where he obtained a lease to a mining property and a place to live in Helen. Shortly after operations began, his contractor friend from Columbia came by. After a hearty exchange of pleasantries, the man asked to borrow $200.[15]

From the start of the Gold Rush, there always seemed to be a few Englishmen around. They also reappeared during the 1930s to operate the Franco-American Mine at the upper end of the Helen valley, where they drove a number of tunnels into Little Hamby Mountain. Leigh Gedney visited this mine, and recalls the proprietors being "very British in speech and dress". They gave him a tour, leading him into a tunnel which soon opened into a large room. The view gave an insight into mining practices, for the roof was supported in an unusual way. In the center of the room, the miners had left a large rock. Upon it sat a car jack, which pressed a large beam against the ceiling.

1930s MINE. Scenes from the Franco-American Mine on Little Hamby Mountain. Rock ore obtained from tunnels above was transported to the mill building in the lower picture for pounding.

With bulldozers, draglines, improved pumps, better explosives, powered drills, dredges and other new technologies, these latter day miners could work some previously difficult areas. Old techniques were not entirely neglected, though, for the Hamby Mining Ditch was also put back into operation. Their tools may have been better, but the results were eventually the same as they had been for the previous century: most found little in the way of profits and their efforts largely ceased by the 1940s. Ironically, it was the operator of the old Hamby Ditch who had the most success. Gold mining was again interrupted by a war, as the federal "War Production Limitation Board" issued an order banning nearly all gold mining in October of 1942.

Except for a minuscule revival when gold prices spiked to about $1000 an ounce around 1980, mining activity since World War Two has been very sporadic and typically an offshoot of sand and gravel operations. While some hope for and even predict another gold rush, it's hard to see how one could occur. The first miners were helped by an unseen hand, for they harvested rich valley placers formed by countless aeons of natural erosive action. To a lesser extent, the later hydraulickers benefited from the same thing as they worked slope placers and the weathered saprolite.

Generations of miners have labored to find and work the "hot spots" in the gold region. This "easy" gold is long gone. Although the rich valley placers suggested a "mother lode" nearby, they turned out to be the handiwork of Mother Nature, who spent millennia concentrating the precious metal on the valley floors. The source vein deposits have simply not proven rich enough or extensive enough to support continued operations. Although much unrecovered gold certainly remains in the area, any new rushes will have to await significant increases in price and/or improved ways of recovering it.

For many outsiders who came with high expectations and left with big losses, the Georgia gold region was "not overly pleasant". But for a few hobbyists and some locals who were not entirely dependent upon it, gold mining has long been an enjoyable and occasionally profitable activity. Even today, armed with only a shovel and gold pan, anyone can work along a pretty creek whose rippling waters flow through a scene illuminated by dappled sunlight filtering through overhanging thickets of laurel and rhododendron, and find some gold there, no matter how many times the stream has been worked before. And, as unknown regions below the surface are explored and earthen smells scent the damp valley air, there is always the enticing possibility of finding something more than gold dust and little flakes, for larger nuggets yet reside in the north Georgia hills.

Modern miners should note that, as has always been the case at the eastern end of the Dahlonega Gold Belt, the institution of private property constrains would-be gold diggers. Even though Helen tourists wielding metal detectors have been known to dig up city water mains in a vain search for the elusive metal, all of the gold lands are owned by someone and permission should be obtained before visiting them. An authentic experience can be had at local "gold and gem" panning places, and there is a local gold mining club for more serious hobbyists.

Like most settlers in the area, those in the Helen valley worked their own lands for gold. The Bells and Englands worked along the creeks and dug the valley placers which underlay their fields. Adam Pitner was very active, selling shares in his placers and investing in many mines elsewhere along the gold belt. On the Conley place, the fields were worked and there are tunnels and a hydraulic cut on the west side of the river.

Even so, the pioneer settlers were not the most active miners in the Helen valley. Richard England's name was applied to the "England Mine" at the south end of the valley, but he was not the one who worked it. And while the largest mining operations in the valley also involved lands once owned by Richard England, the work was done under the guidance of a New England school teacher whose eventual residence in the Helen valley resulted from his troubles with a balky mule.

LIFE AND HARD TIMES AT THE ENGLAND MINE

For several years before the Georgia Gold Rush began, Richard England worked to improve his fields between the Unicoi Turnpike and the Chattahoochee River at the lower end of the Helen valley. Today, these fields are home to Helen's sewage treatment complex (known locally as "Twin Lakes") and the Comfort Inn. And the old England Place is also host to another modern enterprise, the "Gold Mines of Helen".

This popular tourist attraction was established in the early 1990s by Dick Showalter, a promoter of dreams and genuine character who would have fit right in with any of the wild men drawn to the Georgia gold region over the years. Among other things, Dick claimed he had learned of the England Mine from the Seminole Indians down in the Everglades (renegade Cherokees told them about it 200 years ago), that Spanish explorer DeSoto had worked the spot (Dick's least original claim), that it was the site of a Cherokee "cliff dwelling" (them western Indians have nothing on us), and even that Paul Bunyan had once been there (he left his greatly oversized gold pan, which Dick was able to recover). While some criticized his style, he was very entertaining and there would be no historic tourist attraction without Dick Showalter.

The mine has since been purchased by Albert Weisner, but the setup is similar. Modern visitors can

Dick Showalter sitting on bricks which once supported the boiler for the steam-powered 10 stamp mill at the Plattsburg Mine. This site was originally known as the England Mine.

take a damp walk back into an old mine shaft or pan for gold and semi-precious gemstones. There is a collection of old mining paraphernalia and a video presentation on mining history. The current venture has been successful so far, but that has rarely been the case with the old mine.

Paul Bunyan and cliff dwellings notwithstanding, the story of the old mine actually starts about ten years after the Cherokees vacated the area east of the Chestatee River. After the Helen valley was opened to white settlers, Richard England purchased his 250 acre Land Lot in 1826 for $1500. By the time of the Gold Rush he had his farm in good shape. Although it wasn't yet a gold mine, it was productive enough so that he could buy an adjoining 40 acres of land and pay for it with 55 spare bushels of corn.[1]

By the summer of 1831, John Humphries had worked his way up the Chattahoochee to figure out there were significant deposits of placer gold in Richard's lower fields, the source of which appeared to be quartz veins in the steep, rocky ridges across the river. A deal was soon struck whereby Humphries agreed to pay $10,000 for half interest in this part of the England property, giving Richard $5500 in cash and a mortgage note for the remaining $4500. Although John Humphries was a former sheriff of Habersham County, he'd lately had his share of troubles with the law, having been charged with "false imprisonment" in 1830 and with "obstructing a constable in the execution of his office" a year later.[2] And as things turned out, these were not to be his last encounters with the law.

Three other men soon joined with Humphries to form a mining company, paying Richard England a total of $9000 for the remaining half interest in the 100 acres which became known as the "England Mine". A ditch was dug through the middle of Richard's fields to carry water for placer mining operations. Hard rock mining began on the original "England Gold Vein". A dam was built across the river, and a stamp mill set up to crush the ore from the mine.[3]

After mining operations began, Richard England used some of his newfound wealth to acquire a more favorable homesite about a mile up the river, moving his family to the upper end of the Helen valley. John Humphries took over the vacated England home, giving Richard another mortgage note for his old house and the surrounding 3/4 acre.

In 1832, Archibald McLaughlin arrived on the scene. He had been hired by Daniel Blake, a prosperous Savannah planter, to make investments in gold properties. For $2500, McLaughlin bought a 1/16th interest from one of Humphries' partners and transferred it to Blake.

The Blakes had already profited from New-World gold, and did so without actually having to mine it. The Blake family fortune derived from an Admiral Blake who commanded the British fleet in the 1600s, preying upon Spanish ships bringing gold and other valuables from the new world. The Admiral got to keep part of the loot and gave a small share to his brother Benjamin Blake, Daniel Blake's direct ancestor.[4]

EARLY STAMP MILL. Stamp mills could be made almost entirely of wood, except for the iron bands or plates attached to the bottoms of the heavy stamps. Designed and built by a resourceful local man, this mill was photographed about 1895. It has three rough-hewn stamps, which were repeatedly lifted and released to crash down on the ore below. The small sluice box in front was used to separate and catch the fine gold after the crushing was done. Later mills of steel and iron had twenty or more stamps; some of these were steam driven. A working replica of this early mill is at Gold Mines of Helen.

Thus enriched, Benjamin Blake came to South Carolina in 1683 to escape religious persecution. The next five generations of Blakes owned large plantations in South Carolina and many held high offices in the state, but all were educated at the finest schools in England and many were born there.[5]

Daniel Blake was born in England in 1775 and educated at Cambridge. After marrying in 1800, he moved first to South Carolina before settling in the Savannah area. Blake's fine upbringing seems to have prepared him favorably for the gentlemanly enterprise of managing a tidewater plantation, but he didn't fare quite so well in the wild Georgia gold region.

Blake's first trouble was with Archibald McLaughlin. They had signed an agreement whereby McLaughlin would trade only for Blake's account and no other. But McLaughlin began buying property for himself all along the gold belt, and apparently some of Mr. Blake's money got mixed up with his. When Blake found out, he brought charges against McLaughlin for "cheating and swindling" and their relationship was dissolved.[6]

However, Blake was not finished with the England Mine. When he saw the place, it was looking like a gold mine should, with the dam and the ditch feeding the placers and the vein yielding a steady supply of ore for pounding in the stamp mill. And, as was typical in the first workings of many north Georgia mines, the placers were probably showing good returns. In the fall of 1832, Blake bought out Humphries' three partners, paying them more than twice what they had paid and bringing his total investment to over $20,000.

For the next two years, the patrician Mr. Blake and tough ex-sheriff Humphries were equal partners in the England Mine. Both men had a strong case of gold fever, for each made numerous investments in other gold mines during this time. Things were easier for the wealthy Blake, who also had financing back in Savannah and brought considerable cash to pay for his mining interests. Lacking such resources, Humphries mortgaged and borrowed his way through the gold region, putting himself in a position where he had to either find plenty of gold or go broke.

With at least 75 slaves in the gold region, Blake probably assigned some to work the England Mine. Humphries also owned slaves, two of which he put up as collateral for a $600 loan after work began at the mine. However, even with the large expenditures of money and labor, the first operations there were short lived.

By the winter of 1834, John Humphries had left the Helen valley and moved about thirty miles west to Lumpkin County, where he won a contract to build that young county a courthouse in the new city of Dahlonega. He was low bidder at $7000, promising to complete the building in 18 months. County officials advanced Humphries $2500 to get the project started.[7] Humphries was also a partner in the newly organized "Belfast Mining Company" there. As one of the first big-capital mining companies chartered by the state legislature, Belfast was authorized to issue $500,000 in capital stock.[8]

In December of 1834, Daniel Blake died in Savannah. His friend and associate Robert Habersham was administrator of his estate. Habersham found that Blake had, along with his numerous mining properties, a number of negro slaves and "other personal effects" scattered all over the gold region, with much of this property likely beyond recovery. Habersham also found that the titles to the mines were "in many cases wanting or not to be found among the papers and in others the claim of title appears to be defective."[9]

By 1835, the wheels had come off of John Humphries' mining operations. The England Vein had not produced as expected and none of his other investments were paying well enough to keep him afloat. As his creditors lost patience, Humphries' gold properties began to be seized to pay his debts. Back in the Helen valley, Richard England brought suit against Humphries for the unpaid $4500 mortgage he had taken four years before, forcing Humphries' half interest in the England Mine into public sale on the steps of the Habersham County Courthouse. There, it was purchased by Daniel Blake's two sons (Daniel Jr. and Arthur Middleton) for $2501, giving the Blakes total ownership of the property. And even though the proceeds from the sale were several thousand dollars less than Humphries had originally promised, Richard England had still done pretty well. With this final collection, he had realized about $17,000 from the sale of the mine.

If things were bad for Humphries in Habersham County, they were worse in Lumpkin County. There were no steps on the courthouse there or even a building to step up to, since the 18 months allotted for its construction had passed and Humphries had done no work at all on the new structure. Somewhat belatedly, county officials sent the law after him. All they got back was a report, for the results were as follows:

> In attempting to serve the within writ the defendant, John Humphries, stood in defiance, armed with a pistol, which he drew, so he could not be served and after I had summoned the posse comitatus he had concealed himself and has absconded from the county of Lumpkin. Fifteenth day of October, 1835.
>
> John D. Fields, Jr.,
Deputy Sheriff.[10]

OLD HABERSHAM COUNTY COURTHOUSE. Until White County was carved out of Habersham in 1857, legal and political affairs centered on the Habersham County Courthouse in Clarkesville. Built in 1832, this was the scene for the public sale of mining interests owned by John Humphries and Daniel Blake, including those in the England Mine. The mine later put in several public appearances at the White County Courthouse in Cleveland.

Back in Habersham County, the departed Humphries' involvement with the England Mine officially ended in March of 1836. Humphries had never paid the second note he gave Richard England for his house and 3/4 acre several years before. Richard had died by this time, but his widow Martha and oldest son Athan forced the property into public auction at the Habersham County Courthouse, where they were high bidders, buying their old home-place back for $28.25.

Meanwhile, Daniel Jr. and Arthur Blake were making progress in cleaning up some of their father's affairs in the gold region. A number of negroes were located and sold, including "Pompei" who went for $200 and "Ceasar" who sold for $50. They also advertised a public sale of mining properties to be held at the Habersham County courthouse in May of 1836. On the day of the sale it became clear that the elder Blake had lost a fortune in the gold region. One mining interest for which he'd paid $500 was sold for $5 and another for which he had paid $1000 went for $20. The half interest in the England Mine for which Daniel Blake had paid Humphries' partners over

$20,000 was sold for only $300, and would have brought less except it was bought by the two sons and Richard Habersham.

The Blakes, though, apparently could spare the money. Daniel's heirs were in no hurry to dispose of the England Mine or many of their other north Georgia gold properties. The family held their interest in the England Mine for years, as did the descendants of Richard Habersham. The Blakes also kept their interests in other mines, eventually making substantial sums with one or two of them. They never made any money on the England Mine, though, in spite of the many mine shafts and hydraulic cuts which show the site has been worked several times since the days of Daniel Blake and John Humphries.

By the 1880s, the Blake interests in the England Mine had been passed to the grandchildren of Daniel Blake. Daniel Jr.'s interest passed to his eldest son Frederick. And in 1880, the by-then elderly Arthur Middleton Blake simply gave away his interests in a dozen north Georgia gold mines, including his interest in the England Mine. By the deed, Arthur was "of the Charleston District of South Carolina", but in traditional Blake fashion was "now living in England and member of the Reform Club of London". Ownership stayed in the family, though, for the lucky recipient was his nephew of the same name, also Arthur M. Blake, who lived in Tennessee at the time but later moved to lower Georgia.

Considerable work was done several years later by G.W. Sylvester, who in 1882 obtained the younger Arthur Blake's half interest in the England Mine for the nominal sum of five dollars. Tunnels were cut, and it was probably around this time that the mine was hydraulically mined by means of a ditch run around the mountain from the headwaters of Bell Branch. Bell Branch is a small stream, and by most accounts, this ditch was connected to a much longer canal which captured the waters of Smith Creek near Anna Ruby Falls and ran through today's Unicoi State Park.

Sylvester operated the mine for several years, but by 1885, he suffered the same fate as John Humphries had a half-century before. The only difference was that Sylvester's interest was auctioned on a different set of courthouse steps since the mine was by then in White County, a new county which had been carved out of Habersham County. The sale was forced by a labor lien filed by C.P. Craig, who himself wound up buying the half interest for $500. Since the sale was to satisfy his lien, Craig got all of his money back except for about $8 in court costs. Two years later, Craig finally got paid for his labor when Arthur Blake reappeared and paid him $500 to re-acquire the half interest.

Through the rest of the 1880s, ownership of the old mine was still held by Blake and Habersham descendants. Arthur Blake had regained his half, while the other half was still held between his older brother Frederick and William Habersham, a descendant of Richard Habersham. Although these descendants had little apparent interest in the old workings, the England Mine was not forgotten in north Georgia, where gold fever was still making an occasional appearance. In 1890, two local men approached Arthur Blake, who sold each of them some part of his half interest. Arthur Blake then dropped from the picture, but the two local men subsequently appeared in court, apparently to contest each other's claims of ownership. During the proceeding they were joined by an attorney representing Frederick Blake and William Habersham.

In October of 1891, the matter was resolved in a familiar setting when ownership of the mine was consolidated by court order and sold on the courthouse steps. The high bidders were the two local men, who had now apparently decided to work together to recover their investments in the property. But instead of keeping it, the local men transferred their bid to Jay P. Mitchell of Michigan, who for $990 became sole owner of the England Mine.

In late 1895, Mr. Mitchell joined Richard England as a big winner with the mine, selling out to Procter Pettingill and the "Plattsburg Gold Mining and Milling Company" of Plattsburg, New York for $15,000. The high price smacks of gold fever, but the new owners tried to be careful. In early 1896, a visiting state geologist found their prospects favorable, even though he noted the mine had been "purchased at a large price". Before buying, the new owners had sent what they believed to be a representative sample of ore to an assay company in New York City. The resulting report found the ore to be low grade, but of such quality that over 80% of the gold it contained could be caught on amalgamation plates after crushing. Based on this report and the estimated quantity of ore, both the geologist and the company expected profitable operations. Things were soon underway, as a chute was built up the hillside to direct ore to the steam powered ten-stamp mill erected beside the Chattahoochee River.[11]

In May of 1896, an Atlanta Constitution article reported that the Plattsburg stamp mill had been running for about two months, and "The results are so satisfactory that a more extensive plant is being contemplated." Whether this report was based on the usual early optimism or represented a more calculated exaggeration isn't known. However, before the year was out, the new crew was having difficulties, with thirty-two liens filed against them.[12] The next geologist visited the mine in 1909, but he didn't have a whole lot to say. After observing that a lot of work had been done on the old place and noting an abandoned ten-stamp mill beside the river, he wrote that "No mining operations have been in progress for a number of years. . ."[13]

And none have been in progress since then, either. Like many other north Georgia mines, the England Mine has been the scene of much hard work and the final resting place for more than a few broken dreams. Richard England and Jay Mitchell bought low and sold high to make considerable profits, but for everyone else, the mine seems to have been a hole in the ground where they poured money. With the collapse of the New Yorkers' Plattsburg operation, the mine also became an example of the "South's Revenge" extracted by the Georgia gold region.

It's been known since the days of John Humphries and Daniel Blake that gold was waiting to be found in the steep, rocky ridges across the Chattahoochee, a fact which remains true today. The problem, though, is the same as it has been for generations of miners: until somebody finds a big, rich vein which has so far escaped detection, the ore on the site is simply too poor and too hard to get for anyone to make a profit with the mine.

Unless, of course, they try something like the modern tourist operation, which has in a way turned the old mine inside out. History shows that it's a lot better idea to have flocks of the curious bring money in than to try scratching it out of the steep, rocky ridges beside the Chattahoochee River.

OLD ENGLAND VEIN. Employees of the Plattsburg Gold Mining and Milling Company pose in front of operations on the hard rock of the "Old England Vein", probably in 1896. Preliminary assays on ore from the site indicated mining would be profitable, but the company was out of business in less than a year.

* * * * *

THE MATTER OF THE UNACCOUNTED INTERESTS

In 1986, a man buying a large tract of land near Helen hired an attorney to check the titles of the various parcels he was acquiring. One property thus investigated was "Land Lot 91", a 250 acre lot first laid out by Georgia surveyors in 1820 and still intact all those years later. That it was still in one piece was unusual, for most of the old land lots had long since been broken up by the better part of two centuries of real estate dealings. Although it was crossed by a stream which came to be known as Alabama Branch, it was steep and not much good for farming. Old 91 seems never to have attracted a permanent resident, but there was one very special thing about it. It lay along Dukes Creek, at or very near the exact spot where gold was first discovered in 1828 or 1829.

In reviewing the "chain of title", the attorney discovered at least seven "unaccounted" interests in LL 91. His dry legal discussion didn't reflect it, but all of the missing interests had been acquired by men drawn to the Georgia mountains by the lure of gold. Among missing owners on his list were some of the most prominent early miners in the area: Arnold, McGhee, and Hume. Also on the list were Daniel Blake and Richard Habersham.

The situation on Land Lot 91 was not unheard of, for the great Gold Rush scrambled deeds all along the gold belt. In some cases fraud and poor record keeping caused confusion, but this does not appear to have been the story on old 91. Half of LL 91 sold for $70 shortly before gold was found, but when the site turned out to be an auriferous "hot spot" a few months later, the other half quickly sold for $5000. In the frenzied trading which followed, interests in each half were sold, and further subdivided and sold again so that successive buyers eventually obtained title to shares as small as 3/24 of the south half and 11/80 of the north half.

This type of trading often generated profits along the way like a pyramid scheme. But when the fever subsided and a mine played out, those caught holding such interests could see their stakes plummet in value to the point where they could hardly be sold at all. Such a collapse appears to have occurred on LL 91 where the trading abruptly ceased shortly before 1840, probably when the stream-side valley placers along Alabama Branch played out.

As they had done with their interests in the England and other local mines, the first several generations of Blake and Habersham heirs found no advantage in selling, so they held on to their pieces of the LL 91 property. And

As they had done with their interests in the England and other local mines, the first several generations of Blake and Habersham heirs found no advantage in selling, so they held on to their pieces of the LL 91 property. And while circumstances brought about the sale of the England Mine when these heirs could still be identified, nothing similar happened on LL 91. Never reappearing in local deed books, the interests passed on to succeeding and increasingly remote generations for whom any connection to the gold region probably consisted of a few old family tales and perhaps some yellowed papers saying something about ownership of distant and likely worthless gold mines which they had never seen. Like pennies in a parking lot, their interests in the old mines just weren't worth enough to bother picking up.

The attorney stated in his report that a "diligent" search had been made to locate the heirs who owned the unaccounted interests, but none could be found. But since the search was local, there was never any chance of finding the missing owners. The lost interests had all been obtained by outsiders, people who came in from South Carolina, Tennessee, Savannah, and even England, and who left the country when their investments went sour. And for at least a century, none of their descendants had returned to protect or investigate their part ownership of a north Georgia gold mine. The matter of the unaccounted interests was eventually resolved in a judicial proceeding. The court nullified them, for a "lack of interest" as it were, giving full title to the new owner.

The Gold Rush brought thousands of people to north Georgia in the early 1830s, and many more over the next century. West of the Chestatee River, gold-seekers were among the first to live in the region when it was wrested from the Cherokees. But to the east, settlers were already there. Many miners labored in the hills around the Helen valley and Nacoochee, but — like the missing owners of old Land Lot 91 — most of them left. They had an influence, but they did not displace or override the culture of the original mountaineer and piedmont settlers who had already occupied the area.

THE GREAT MINING DITCH OF J.R. DEAN

Although just about everyone who lived in the Helen valley at least dabbled in gold, the foremost resident miner turned out to be a somewhat unlikely character. He was Josiah Robinson Dean, a former New England school teacher who migrated southward before the Civil War in search of better health. Known as J.R., he was about thirty when he left the North. After living for several years in Knoxville, Tennessee, he determined to remove even further south. While making an exploratory trip to northeast Georgia, J.R. found himself stranded as he traveled along the headwaters of the Chattahoochee River. A grandson provided this account of the Dean family odyssey to Georgia:

> By a gradual progression that consumed some ten years, an Anglo-Saxon family began shortly after 1848 an emigration from the bleak winters of New Hampshire, Massachusetts and New York to the sunny summers of the South, finally locating in the northeastern section of Georgia famed in Indian legend and known as Nacoochee Valley.
>
> There, amid the graceful slopes of the Blue Ridge and in a climate salubrious and well suited to return the father's loss of vigorous health, Josiah Robinson Dean, of Oakham, Mass.; Rebecca Cooke Dean, of Ackworth, N.H., his wife, and Clara Dean, daughter, settled upon some mining properties. The poor health of the husband, the reason for his move to the South, caused him to forsake his chosen profession of teaching, which he had pursued for many years and in several leading colleges of the East, and to enter the field of gold mining. The wife continued her teaching career, this time among the scant peoples of the mountains of North Georgia.
>
> Behind them, beneath the sod of Tennessee, they left two babes, Alice and Anna Dean.
>
> Choice of a mining vocation by Mr. Dean was in many respects a modern instance of the Biblical Balaam and the ass. Mr. Dean left his family in Tennessee and astride a mule set forth southward on a tour of investigation to find a suitable place to locate permanently.
>
> Coming to a stream in North Georgia, the creature balked with the characteristic cussedness and stubbornness of the breed, and no pleading, cajolery or pressure could make it take the

water. A mountaineer saw the rider's dilemma and called out to him a friendly greeting. A conversation ensued, the mule remaining on the unwanted side of the stream.

In the course of the conversation, Mr. Dean told of his search, saying that if he could ever inspire the brute to ford the stream he would head towards Clarkesville, where he had heard there was a good year round climate and considerable business opportunities.

The mountaineer, knowing Mr. Dean was seeking health and that outdoor life was what was needed, suggested he purchase a gold mine further down the valley. Mr. Dean became interested, investigated, purchased, and then moved his family down.

The mule never crossed the ford.[1]

Since Mr. Dean was coming down from his first southern home at Knoxville, he was almost certainly travelling the Unicoi Road. And as for the stream crossings, his mule may have simply had enough, for the recalcitrant beast had just made over twenty of them on the way down from Unicoi Gap to reach the valleys of Helen and Nacoochee.

When J.R. first appeared in the Habersham County deed books in 1856 he was still listing his home as Knoxville, but his southward move was completed by the following summer.[2] The 1860 census taker found the Dean family living on the estate of C.L. Williams in Nacoochee Valley, where they appear to have stayed until purchasing a homeplace of their own in 1867. J.R. and wife Rebecca had three more children after coming to Georgia, daughters Etta and Ida, and their only son Herbert Henry, born in the winter of 1861. There were three generations living in the Dean house, for J.R.'s elderly father also followed him to Georgia.

J.R.'s intentions became clear when his name began appearing in 1857 on deeds involving thousands of acres. Hydraulic mining was his aim as he assembled the rights of way and easements necessary for the construction of a great mining ditch to capture the waters of Dukes Creek. This large Chattahoochee tributary runs parallel to the Helen Valley and normally merges with the river down in Nacoochee Valley below the Indian Mound. However, when J.R.'s ditch was done, the normal order of things was changed and the merger of Dukes Creek moved three miles upstream to the Helen valley.

Hamby Mountain is the highest point on a long ridge which divides the lower reaches of Dukes Creek from the main valley of the Chattahoochee. Hamby overlooks the Helen valley from its upper end and today is marked by the WHEL radio tower which stands upon its summit. A State geologist sent to survey Georgia gold mines described the Hamby ridge as follows:

> Topographically considered, this region forms the watershed between Dukes Creek and the Chattahoochee River, and is a long spur of the Blue Ridge. . . . It has one prominent elevation known as Hamby Mountain, which. . . attains an elevation of over 2300 feet above sea-level. This stands in the midst of the Nacoochee mining field [at this time both the Helen valley and that of Dukes Creek were considered part of Nacoochee] and gives rise to many mountain branches, which flow down its sides, and stretch out south and north to Dukes Creek and the Chattahoochee River respectively. . . . Present evidence points to the saprolite as the chief supply of the streams gold to this property, and large areas of it are to be found here. Indeed, one can hardly wash a pan of dirt along any of the prospecting cuts without obtaining some gold. . . . The total amount of gold already taken from the branch deposits is something enormous. The branches, especially during the wet months, are perfect torrents, with an immense erosive power. . . . Centuries of such powerful and continued action on the feeding saprolite and decomposing veins of this auriferous body of land have made the deposits of these branches the most richly productive in Georgia.[3]

Since the valley placers and some veins along both sides of the Hamby divide had been worked extensively for a quarter of a century before J.R. Dean arrived, the richness of the area was well known. Although J.R.'s plan was grand in scale, his vision was simple: by carrying the necessary waters hundreds of feet higher up the mountain, his ditch would open for mining thousands of untouched acres along the Hamby ridge which likely contained much more of the gold which had fed the rich valley placers down below.

In 1857, the "Nacoochee Hydraulic Mining Company" was chartered by the State. Among the partners were J.R. Dean, George Cook of Amehurst, Massachusetts and Collett Leaventhorpe, an Englishman who came to Georgia by way of Rutherford, NC. J.R. maintained his New England connections, for a series of Boston area investors provided much of the financial backing necessary to carry out the grand plan he brought to north Georgia.

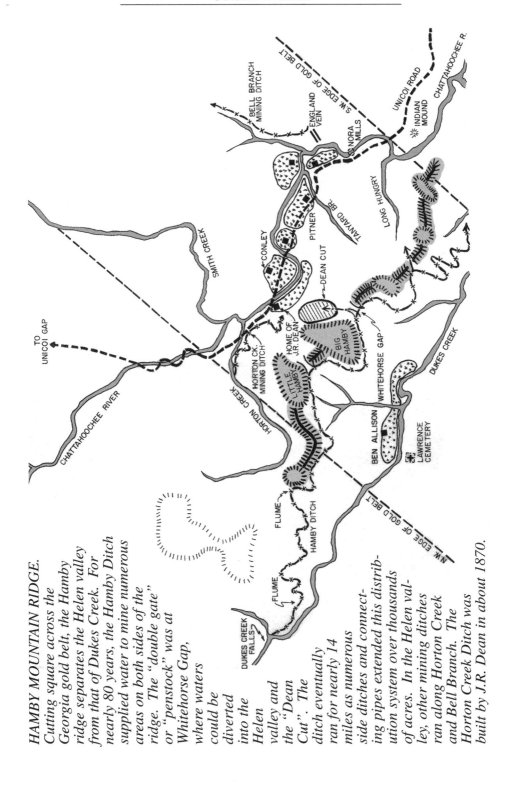

HAMBY MOUNTAIN RIDGE. Cutting square across the Georgia gold belt, the Hamby ridge separates the Helen valley from that of Dukes Creek. For nearly 80 years, the Hamby Ditch supplied water to mine numerous areas on both sides of the ridge. The "double gate" or "penstock" was at Whitehorse Gap, where waters could be diverted into the Helen valley and the "Dean Cut". The ditch eventually ran for nearly 14 miles as numerous side ditches and connecting pipes extended this distribution system over thousands of acres. In the Helen valley, other mining ditches ran along Horton Creek and Bell Branch. The Horton Creek Ditch was built by J.R. Dean in about 1870.

As J.R. acquired mining rights to hundreds of acres along the Hamby ridge, in most cases the properties were not purchased outright. Instead, leases were obtained in exchange for a promise to give the owner a fraction of any gold found on his property, usually a 1/10th share. Land dealings in the gold region were complex, particularly with a project the size of the Hamby Ditch. J.R. soon became entangled in enough legal disputes to establish a long and regular relationship with prominent Gainesville attorney Col H.H. Perry.[4]

The plan from the start was to spill the waters of Dukes Creek on both sides of the Hamby divide. Above Hamby Mountain, the ridge was too high, but at Whitehorse Gap on the lower end of the mountain and at several other points on down the ridge, the transfer could be accomplished. By 1860, J.R. had selected the Helen side of Hamby Mountain as one of his intended mining areas. In that year, J.R. negotiated a ten year lease with Martha, Jerome

THE DOUBLEGATE. At the lower end of Hamby Mountain, the Hamby Ditch was high enough to send the waters of Dukes Creek through Whitehorse Gap and into the Helen valley, where they made an unnatural entrance into the Chattahoochee three miles above their normal merger. The waters could also be sent on down the Hamby ridge through many more miles of ditch used to work areas on both sides of the divide.

and Coleman England, all of whom still lived on the old "Widow England Place" at the base of the mountain.

Since the England lands included both the fields and adjacent slopes at the foot of Hamby, they would be affected by waters brought through Whitehorse Gap to the Helen side of the peak. But, for a 1/8 share, the Englands were willing to allow J.R. "to mine for gold by the hydraulic hose process of mining" on and above their homeplace. However, since such mining was a messy process, the deed contained a covenant to protect the England fields:

> Dean shall secure our lands from damage by carrying water through our bottom lands in a box extending from 50 yards above our fence to the Chattahoochee R. and pay for any damage which may be accidentally done.[5]

When the necessary capital and property rights were secured, the company moved quickly to construct what became known as the "Hamby Ditch". It was an expensive undertaking for its day. However, as noted in a State geologist's report, the Civil War halted operations before they could have much effect on the England farm or any place else:

> . . . at the breaking out of the war, the present canal was completed, water running through, and ready to be used for the hydraulic operation of the gold fields, which it commanded. The cost of building this canal was about $50,000. The war suddenly arrested all work; but upon the restoration of peace, work was recommenced. . .[6]

Once work was restarted, the Hamby ditch remained in regular use through the early 1900s. The ditch captured the waters of Dukes Creek just below Dukes Creek Falls. Although this intake was well outside the gold belt, having the pickup so far upstream was necessary to achieve the maximum height for the waters when they reached the mining areas below. Flumes were required along the upper reaches. While noting that the ditch reached a height of 300 feet above the surrounding valleys, the State geologist provided a detailed description:

> This aqueduct. . . winds around slopes of Hamby mountain with a gentle and uniform grade, [with a fall of] five feet to the mile. Its course must necessarily be uneven; but it is rendered moreso by the engineer's endeavor to avoid the building of trestles [flumes]. . . . Although the lower end of the ditch is only three miles, in a direct line, from the dam, its complete length is about seven and one-half miles. Its elevation above,

and its location in respect to, the known deposits is such that it is universally acknowledged to be the key to the entire auriferous area between Dukes Creek and the Chattahoochee River.[7]

START OF THE HAMBY DITCH. The intake for the Hamby Ditch was just below Dukes Creek Falls. In use by the Civil War, the original ditch was 7.5 miles in length and attained a height hundreds of feet above the surrounding valleys when it reached Hamby Mountain. Rebuilt at least twice, the ditch nearly doubled in length before operations finally ceased in the late 1930s.

Shortly after the war, in 1867, J.R. moved his family to the Helen valley. He had obtained only a mining lease to the Widow England Place seven years before, but after Martha and Coleman England died, J.R. was able to buy it outright. The seller was Jerome England, the last of the pioneer Englands to live in the Helen valley.

The England house was at the lower end of their fields, close beside a wagon road which crossed Hamby Mountain at Whitehorse Gap.[8] Although the old cabin was presumably still standing, J.R. decided not to live there, perhaps because it was already 50 years old and was likely only a crude cabin to start with. Instead, the Dean family occupied a new house at the very upper end of the England fields, one which sat on a small rise overlooking the upper ford of the Unicoi Road.[9] Long-time area resident Comer Vandiver remembers this house as being built of stacked logs with weatherboarding applied to cover the gaps between them.

The Hamby Ditch was inspected a few years after the war by Dr. Matthew Stephenson, who published an account of it in 1871. Although best known for his later activities in Dahlonega, Stephenson lived first in Nacoochee, where in 1834 he acquired interests in a number of local mines for $5000. In spite of a reputation for extravagant statements, Dr. Stephenson had long been recognized as an expert on the Georgia gold belt. Some of both qualities show in his description of the Hamby Ditch (which he refers to as the "Nacoochee Canal"):

> In White and Lumpkin Counties the California hydraulic process is now used for washing gold. The canal which conveys the water to the placers of Nacoochee is seventeen miles long, and terminates at an altitude of three hundred feet, giving the miners every necessary facility and power to wash off the hills and surface grounds for miles around. . . . The power obtained by the hydraulic process, with a head of water of one hundred feet, is estimated, by practical miners to be equal, in washing off the surface earth, to thirty times that of one hand without extra power. The water is brought from the canal, down the slopes of the hill, to the gold placers, and issues from the nozzle of a pipe from one to two inches in diameter, the force of which dissolves the clay and partially decomposed slates [saprolite] with magical rapidity. Accidentally striking a man or animal, they would be killed in an instant. The works at the Nacoochee Canal are paying good profits to the company, who live mostly in Boston.[10]

Although Dr. Stephenson was impressed enough with J.R.'s canal to somehow add several miles to its actual length, he was not entirely pleased with other aspects of the operation:

> Even they [J.R. and company] with their boasted intelligence and wealth, have made no attempt to work the rich veins they continue to find by surface washing, and which have yielded,

ready for the mills, without cost, thousands of tons of better ore than any of the numerous mines in California or Colorado.[11]

The Doctor attributed this "culpable apathy" in neglecting the veins to the "false notions" held by Northerners that the South was "unsafe for any one to live in, or to do business in, who does not belong to the rebel element of the State". Of course, this assertion seems to overlook J.R. Dean's comfortable twenty year residence in the South and the fact that J.R. was doing his best to recover any gold he could find. Dr. Stephenson's comments also seem to reflect something of the local sentiment towards Yankees, mixed with some hostility towards western mines.

But then, such strongly expressed sentiments were not unusual for Dr. Stephenson. In fact, a few of his words were of such character as to achieve the status of a nationally known popular saying which still resonates over 150 years later. After leaving Nacoochee, Dr. Stephenson next lived in Dahlonega where he was assayer at the U.S. Mint. When news of the California gold discoveries reached the Georgia hills in 1849, a number of local miners worked themselves into a state of excitement and began planning a westward expedition.

Even though Dr. Stephenson knew little about western gold, for miners to abandon his beloved Georgia mines was simply not acceptable. After announcing that he wished to make a speech, Stephenson on a Saturday afternoon mounted the steps of the Dahlonega courthouse to inveigh against the folly of going to California. The text of his speech wasn't preserved, but a key phrase was well remembered:

> . . . Colonel Price who was then a printer's boy in the village, remembers how, standing on the courthouse steps, his long-tailed coat flapping in the breeze, the old Doctor [he was 44 at the time] pointed to Crown Mountain, just south of the Mint Building, and cried dramatically: "Boys, there's millions in it!"[12]

Dr. Stephenson's timing was not good, for his words were uttered during the slow period between the depletion of the "easy gold" in the valley placers and the coming of hydraulic mining. Accordingly, his comments were met with derision, for the miners felt that Crown Mountain, like most other areas, had been worked out. In spite of the impassioned speech, many set out in a wagon train for California.

Although they left Dr. Stephenson behind, they took his words along.

As the Georgians struggled west and then began toiling in the California mines, "There's millions in it!" became a inspirational rallying cry similar in spirit to a "rebel yell". Other California miners took up the phrase, which finally caught the ear of Mark Twain. The noted writer eventually put a version of the expression in the mouth of a character who became "Beriah Sellers" in his first novel *The Gilded Age,* and enduring fame was thereby achieved.[13] The end of the story has it that Dr. Stephenson's earnest words were locally corrupted into "Thars' gold in them thar' hills!", a phrase which will eventually grace the ears of anyone who spends enough time in the Georgia gold region.

Having survived both the war and Dr. Stephenson's scathing inspection, the Nacoochee Hydraulic Mining Company continued operations with the Hamby Ditch until about 1877. One of the sites J.R. concentrated on after the war lay above his new lands at the upper end of the Helen valley. Taking in a good portion of the Helen side of Hamby Mountain, it was worked with waters carried through Whitehorse Gap and became known as the "Dean Cut". For years, water cannons hurled the displaced waters of Dukes Creek against the face of Hamby, freeing countless tons of soil, stones and dismembered saprolite and washing them down to be milled and sluiced below. J.R. appears to have picked a good spot, for these were the comments of the State geologist on a part of the Dean Cut:

> Here, Mr. Dean . . . operated, for several hundred feet, a quartz vein, or rather stringers of quartz, widening from a mere band to eighteen inches, and continuing in a well-defined course; and from the milling of the ore and the ground sluicing of its walls, he is reported to have reaped a rich harvest.[14]

Although J.R. apparently continued to demonstrate "culpable apathy" by failing to follow quartz veins into the hard rock of Hamby Mountain, tunneling really wasn't necessary. Hydraulic mining was much easier, and such opportunities were abundant along the Hamby ridge.

In the late 1870s, J.R. and his partners appear to have sold out to a successor company chartered in the state of New York and known as "The Nacoochee Gold Mining Company". J.R. had shares in this firm, which eventually hired him as superintendent for five months in the winter of 1880. They might have done better to hire him sooner, though, for something had gone wrong for the New Yorkers. When they failed to pay J.R. for his labor, he soon joined over a dozen other local men in filing claims against them.[15] By the time the last of the claims were eventually settled in 1885, the company had disappeared from the scene, a few more Yankees who crashed in the "South's Revenge" along the Georgia gold belt.

THE DEAN CUT. View looking down the Dean Cut during operations conducted by the St. George Mining Company in the late 1800s. The mill building and sluices are in the lower part of the cut. Although this area is now covered with forest, rugged terrain still attests to the years of mining operations conducted on the face of Hamby Mountain.

After a nearly thirty year career as a north Georgia goldminer, J.R. Dean died in 1884. Several years later, in 1887, J.R.'s mining interests were sold to John Martin for $30,000. Martin was a Scotsman who came to north Georgia by way of London, England. Although he came to mine gold, he wound up marrying a local girl (a granddaughter of Henry Highland Conley who had grown up in the Helen valley) and, unlike most other miners, becoming a permanent resident of the area. Where J.R. had been the dominant figure involved with the Hamby Ditch for nearly thirty years, Martin assumed much the same role for most of the next twenty. During this period, the ditch was completely rebuilt and other miners allowed to use its waters in return for a share of the gold.

In 1886, there was renewed interest in the old Dean Cut on the face of Hamby Mountain. The St. George Mining Company sent a mining engineer to evaluate the site, who filed an enthusiastic report. The proprietors of this company were associates of Martin, for they listed their place of business at the same London address as he did. When the State geologist came by in 1896, he found the St. George people using a ten-stamp mill powered with water from the Hamby Ditch.

Working high on the mountain and not far enough below the ditch to obtain adequate pressure for the water cannons, they were enlarging the cut by "the simple erosion of water flowing down from the Hamby ditch". A half dozen men were at work in the cut, "with pick and shovel, guiding and assisting the action of the water. . ."[16] to loosen the soil and saprolite and the quartz it contained, which then washed down the mountain to the mill and sluices below.

When the next State geologist came by in 1909 to visit the St. George property, he noted ". . . a large excavation, known as the Dean Cut. . .", and reported that, "No operations have been carried on here for a number of years. . ."[17] Of course, he was saying this about a lot of old mines, since operations had ceased at nearly all of them around the turn of century. The Hamby Ditch was no longer in service, and he could find only a few placer operations in the area and one or two stamp mills still in use.

The Hamby Ditch was not quite done, though, for it was restored one last time during that last burst of activity along the Georgia gold belt in the 1930s. Under the direction of W.G. Hudson, the flumes were rebuilt and the waters of Dukes Creek transported even further than before along the Hamby Ridge as the ditch reached a length of nearly 14 miles. Some of Hudson's flumes were still standing in the 1970s. There was also renewed activity in the Dean Cut, as the Eastwood Mining Company began burrowing into the side of Hamby Mountain, finally overcoming decades of "culpable apathy" to pursue the sort of vein mining which Dr. Stephenson had advocated years before. Eastwood soon proved that apathy had been the right approach all along as they moved tons of

GOLDEN EGG. W.C. Hudson holding a 112 pennyweight (5 1/2 oz.) gold nugget found on Dukes Creek in May, 1936. The nugget is rounded and stream-worn, indicating it has been transported some distance from the source vein. Restoring the Hamby Ditch in the 1930's for its final years of use, Hudson found a considerable amount of gold.

OPEN PIT MINE. 1930s open pit operation in the Dean Cut on Hamby Mountain. With better equipment, these latter day miners could work areas bypassed by the St. George Company and J.R. Dean.

rock with little result. In contrast, Hudson employed the same approach as J.R. Dean with good returns, applying the waters of the Hamby Ditch until WWII forced him to cease operations.

THE DEAN MIGRATION CONTINUES

Like the geologist's comments about his "rich harvest" in the Dean Cut, J.R.s estate inventory gave another indication that his mining operations were successful. Upon his death in 1884, the appraisers estimated his "Lands, personal property, notes, accts., cash on hand and gold" to be worth $12,544.95, an amount over five times greater than the estates of his neighbors John Conley and Priar Pitner. However, this appraisal turned out to be very low since J.R.'s mining interests were valued at less than $1500, but sold to John Martin for $30,000 just two years later.

This is no reflection on the local appraisers, though, for from their seasoned perspective the mines were only worth the lesser amount. And in the gold region, land prices fluctuated like jackpots at a casino, needing only an outsider with deep pockets and a case of gold fever to reach amazing highs.

J.R.'s inventory was clearly that of a big-league gold miner, for in addition to his interests in nearly 20 mining properties, among the items listed were gold dust, gold scales, quicksilver, mining books and receipts, various pumps, pipes, nozzles, and fittings, surveying instruments and compass, a New York bank account, shares of stock in and a judgement against the Nacoochee Gold Mining Company, and thousands of dollars in notes receivable for the previous sale of mining interests.

Also in the inventory were a few items not found elsewhere in the Helen valley. The Dean place was brightened by music, for the inventory listed "one old organ" valued at $25 and "1 old piano" at $10. One can imagine the notes of an old melody issuing forth from the Dean place, and from there floating softly on quiet mountain breezes to catch the ear of a lonely traveler headed south along the Unicoi Road. And if this wilderness sojourner paused beside the Dean place to savor the pleasant tones, he could enjoy a vista which included the upper ford in the Helen valley — where he would make his 26th stream crossing since leaving Unicoi Gap — and the spreading green fields of the Deans and the Conleys, all set against the ragged edges of a still-virgin forest and the gurgling sounds of the rushing Chattahoochee.

As for life in the Dean home, the same grandson who recounted the family odyssey to Georgia gave a friendly description of J.R. and Rebecca Dean:

> Those who knew Mrs. Dean describe her as a bright, happy very intellectual woman who loved her home and never tired in serving, both there and in her community. She was particularly active in church work and charitable enterprises and she succeeded in passing this trait on to each child. The only school was the one she taught herself. . . . Mr. Dean was a silent, thoughtful kind of man, strict but patient, strong but gentle, loving but at times severe.[18]

In addition to teaching her own, Mrs. Dean taught other children during her years in the Helen valley. One of her students was a mountaineer named Henry Newton Abernathy, a son of the Joel Abernathy who worked and probably lived on the Conley place across the river. Henry was always close to the Deans. He grew up with their only son H. H. Dean. The two remained life-long friends even though they took very different paths in life. Henry Abernathy also lived in a small house beside the Dean home when he first married.

THE DEAN FAMILY IN ABOUT 1866. Looking somewhat more like a Confederate general than might be expected, native New-Englander J.R. Dean was fifty years old when the picture was taken. He died in 1884 at age 68. The small boy in front is Herbert Henry Dean, known locally as "Hub". The oldest girl is Clara Dean, who died in 1880 before the Deans left the Helen valley. The youngest girl is Mary Dean. Shown on the right, Rebecca Cook Dean became pregnant with their last child a year or so later at about age 44.

As former New England schoolteachers, J.R. and Rebecca Dean were different from their neighbors in background and education. They were also dissimilar in religion, for they were Presbyterians in a land of Methodists and Baptists. The Deans were initially associated with nearby Nacoochee Methodist Church, where J.R.'s father was buried upon his death in 1871. In that same year, the first local Presbyterian church was built in Nacoochee valley by Captain J. H. Nichols, a wealthy Civil War veteran who moved in from middle Georgia after the conflict.[19]

The Deans appear to have attended this new church for the remainder of their years in the Helen valley. There is a small cemetery behind this church where, according to an account provided by a daughter of Henry Abernathy, J.R. Dean was buried in 1884.[20] It's also likely that a daughter, twenty five year old Clara Dean, was buried there upon her untimely death in 1880. However, neither is buried there now.

After J.R.'s death, wife Rebecca went to live with a daughter about thirty miles away in the city of Gainesville. Their only son Henry Herbert remained on the old Dean Place for about a year before departing to spend the next year in law school at Athens. H.H. then joined his mother and sister in Gainesville, where he began a very successful legal practice with the help of the Col. Perry who had long handled his father's legal affairs.[21]

When Mrs. Rebecca Dean died in 1897, she was buried in Alta Vista Cemetery, which is the Gainesville city cemetery. Alta Vista sits just across the road from the former Gainesville home of H.H. Dean. Buried beside her are J.R. and their daughter Clara. Alta Vista was established in the early 1870s.[22] The Deans are in Block 3, which was opened some years later.[23] The timing of daughter Clara and J.R.'s passing, the subsequent Dean moves to Gainesville, the opening of Block 3 in Alta Vista, and of Mrs. Dean's passing combine to support the notion that the remains of Clara and J.R. were moved to Gainesville by H.H. Dean.

The Deans were very conscious of their ancestry and seem to have been quite proud of the "pure blood" which they determined to flow in their veins. Although J.R. finished his days in the Helen valley, after a thirty year delay caused by a balky mule, the Deans moved on, apparently even to the point of taking some of their earthly remains with them. They seem to have co-existed quite well with their neighbors, but never really mixed with the "scant peoples" of the mountains. In a very basic sense, they came in different and left the same way.

But even though H.H. Dean moved down to Gainesville to become a "high-powered" and wealthy lawyer, he never lost his connection to the mountains. He sold the Dean mining interests after his father died, but eventually re-acquired many of them, usually at bargain prices when later operators ran into difficulty. In fact, the Deans held rights in the Hamby Ditch until 1946, when H.H.'s children gave them to three local men, Lat, M.C. and Comer Vandiver. Some of these rights are still held by Lat Vandiver's grandchildren.

H.H. and his family spent considerable time in the Helen valley, for H.H. eventually built a large summer home at the foot of Hamby Mountain, just below the old Dean Cut where he and his father had worked when he was a boy. H.H. was on the board of the Gainesville and Northwestern Railroad, which could deliver him nearly to the front door of his summer place after the tracks were laid across the old Dean fields in 1913. Today, the Dean summer house is owned by the Greear family, who obtained it from the Deans after H.H. died in the late 1920s.

Mrs. Dean apparently succeeded in instilling the virtue of charity in her son, for Herbert Henry Dean gave generously to a number of causes, particularly those involving education. One local beneficiary was the Nacoochee Institute, a Presbyterian school founded at the lower end of Nacoochee Valley in the early 1900s. H.H. could also be charitable with friends like Henry Abernathy, who rarely left H.H.'s home without a little money even if he didn't want it. The usual pattern was for H.H. (Henry called him "Hub") to offer cash to the visiting Henry, who would then say "I'm a' makin' it alright!" in vigorous refusal. But often, when Henry got home, he would find the money secreted in one of his pockets, whereupon he would display great aggravation at the antics of his city friend.

The relationship between the Dean and Abernathy families continued for many years. H.H. Dean valued his friendship with Henry Newton Abernathy to the last. When H.H. died in 1927, this "tall, raw-boned mountaineer" was one of the few people his widow would see. And when the Deans finally sold their Helen valley property, the acre where their old house had stood was conveyed to the family of Henry Abernathy. Over one hundred years after the Deans left the valley, a daughter of Henry Abernathy still lived on the site of the old Dean house. As the years went by, she did not know the Dean descendants and they did not know of her. But the memory of the ancestral Deans and their kindness towards her family was kept with her as long as she was in residence at the foot of Hamby Mountain.

This daughter was Laura Abernathy Cannon. Laura carried with her not only knowledge of the Deans, but also of the old days, when life in the mountains was much different, and not greatly unlike that of the first pioneers to reach the headwaters of the Chattahoochee.

THE VIEW FROM HAMBY MOUNTAIN

Until recently, the summit of Hamby Mountain was not a place many people went. It's not that it was a large mountain, but it was high enough and its laurel thickets dense enough to withstand most folk's curiosity about the view from the top. The peak takes its name from Needham Hamby, a gold-miner who lived on its back side for a few years during the early 1830s.

Hamby Mountain was part of a large parcel of land assembled a few years ago by wealthy businessman Charles Smithgall, whose announced intention was to conserve and restore a part of the north Georgia mountains threatened by development. In this he appears to have succeeded by selling his lands to the State, which now manages them as "Dukes Creek Woods". Stretching from Helen to the National Forest, this large preserve takes in much of the Dukes Creek valley and many old gold mines along the way.

Smithgall had a system of roads constructed while he owned the property. One road ran across the summit of Hamby Mountain, where a huge radio tower was erected. Although such development seems incompatible with the general notion of conservation, the modern avenue at least makes the view from Hamby Mountain much easier to obtain. Since the road is on the preserve, it can be traveled only with permission.

This new road crosses the old Hamby Mining Ditch in three places, the first being on its way up the north side of the mountain. From this point, there is a good view towards the upper reaches of Dukes Creek and Dukes Creek Falls where the old ditch originated. The sight shows that J.R. Dean was a good engineer, for it's almost impossible to figure along which of the serpentine ridges and myriad stacks of hills the Hamby Ditch runs. Like many good ideas, the notion of the ditch is simple once you see it, but it took something extra to think of it in the first place.

The view from Hamby is worth the trip. With Yonah Mountain soaring behind, the Dukes Creek valley looks much deeper than expected. The gold fields commanded by the waters of the Hamby Ditch lie below and they are immense, thousands of acres stretching from Dukes Creek to the Helen valley and on down the Hamby ridge to Nacoochee. Standing on the summit, there is gold underfoot, for the part of the mountain above the old ditch has never known the miner's waters.

From the summit, the road comes down the Helen side of Hamby, descending a lengthy ridge before turning to run along the Helen city limits at the base of the mountain. At this point the road runs just above the old Horton

Creek Mining Ditch, another canal constructed by J.R. Dean about 1870. This low ditch transported the waters of Horton Creek into the Helen valley for use in washing the valley placers at the foot of Hamby.

Before climbing the mountain again, the road traverses the old Dean Cut. After all these years, it still takes only a glance to see that some major work has been done here. Deep hydraulic gashes scar the hillside and up above, tunnels are cut into the mountain. To compound the insult, this area was used as the city dump in Helen's early days.

Just past the Dean Cut, the road starts its climb to Whitehorse Gap on Hamby's lower side. On the way, the road crosses the section of the mining ditch which conveyed the waters applied to the Dean Cut. Considering the size of the Cut, this ditch is surprisingly small. It is now so completely overgrown with laurel that crawling is the only way to navigate much of it.

At Whitehorse Gap, the road again crosses the main body of the ditch. The "double gate" or "penstock" was at the gap; this was a junction where the captured waters of Dukes Creek could be sent either around the Helen side of Hamby towards the Dean Cut or further on down the ridge towards Nacoochee.

At the gap, the main ditch is easier to walk, for there it is large and not so overgrown with laurel. A stroll along the ditch shows that its waters were applied in many places. Side ditches run out along every flanking ridge it crosses and there are many breaches where old gates once released the waters into waiting pipes or simply to wash down the intersecting valleys.

From Whitehorse Gap, the road descends to the floor of the Dukes Creek valley. After it crosses the creek to intersect another road leading back up the valley, yet another mining ditch is encountered. The large Yonah Ditch, like the Horton Creek Ditch, was a low canal used to wash valley placers. The old deed records indicate that J.R. Dean was involved with this ditch as well.

Proceeding up the Dukes Creek valley, piles of stones and misplaced mounds of earth are cast about as if a titanic struggle has taken place. And this has been the case, as great sections of the valley floor and the hillsides above have been rent and shoved aside by generations of miners armed with everything from picks and shovels to hydraulic giants, draglines and dynamite. Although there were winners and losers, miners prevailed in many battles and much gold has been recovered in this richest of areas at the eastern end of the Dahlonega Gold Belt.

HYDRAULIC WASTELAND. The impacts of hydraulic mining left areas so barren that early film makers came to the north Georgia mountains to use them as backdrops for western movies. Today, however, these areas are hard to find. Although the landforms are about as the miners left them, most are completely covered with mature trees and other elements of the mountain forest. This scene is on Dukes Creek. Mt. Yonah rises in the background.

It's more than a little ironic that Mr. Smithgall chose this area to protect and restore. Hydraulic mining cuts like the ones made here were once so stark that early film makers came to the Georgia mountains to use them as settings for western movies. Countless tons of topsoil have been washed away and the waters of Dukes Creek often diverted to leave much of its course nearly dry for years at a time.

In the Helen valley, the environmental story is much the same at the England Mine and the Dean Cut, and in many lesser spots along the small creeks and hollows which drain into the valley from either side.

The natural order is certainly changed in the old mining areas, but it may be cause for optimism that Mother Nature has shown strong powers of recovery. With the rearrangement of landforms, it will take centuries of natural erosion before signs of the miner's diggings are completely erased. And since some plants thrive in disturbed areas while others cannot live there at all, a naturalist doing comparative studies could almost certainly show that some original species are either missing or greatly diminished in the old workings.

But everywhere, the old mines are at least covered with all the elements of the forest: large trees tower above an understory thick with laurel, while smaller plants thrive in the humus below. The impact of mining was extreme, but once the wounds were inflicted, the hills did not wash away. In contrast, modern development is more widespread and likely to be more permanent in its effects.

There are few areas in the mountains which have not been impacted by man in some degree. Since the first settlers arrived, bears and even the white-tailed deer have been threatened to the point of near extinction not only in the gold region, but all across north Georgia. But like the torn and barren surfaces of the old mines, the bears and the deer have made a comeback in recent years, and it's a good thing to see. What was once a great wilderness is not so any longer, but wildness yet remains.

TALES OF RED STARS AND REAL CARROT GOLD
AT THE PANNING PLACE

A beautiful summer day in the mountains had us looking for a little adventure. We decided to go to the gold and gem panning place located beside the old Richardson house. The operation is located on Dukes Creek, just off of Highway 75 about two miles below Helen. A lot of gold has been found on the Richardson place over the years, for it was the site of some of the richest mining operations ever conducted along the Georgia gold belt.

We got a bucket full of "gold ore" for a few dollars. The sandy mixture was said to come from the creek which runs beside the Richardson house. At the panning affair, a goldminer's pan and large troughs or "water tables" provide the means to process the ore. We filled our pan and began to work it down.

The procedure is to pick out the larger rocks while sloshing and spinning all but a few spoons full of the heaviest stuff from the pan. After this was done, swirling the leavings with a little bit of water exposed the heavy black sand and three gleaming little flakes of gold.

"It's real carrot gold!", exclaimed young Mr. Graham, five-years-old and full of excitement. After hearing that carrots had something to do with gold, he figured out on his own that 18 karat gold and 24 karats were o.k., but "real carrot gold" was the best, and that of course was what he'd found.

According to the latest speculations of modern physicists, in the beginning — or at least right after the beginning, which is as far as the theories go — there was hydrogen, the lightest element. Hydrogen is the initial high-octane fuel of stars, which release tremendous amounts of energy through the process of nuclear fusion as heavier elements are created.

As stars age, they go through many phases, changing in size, color, and temperature, creating ever heavier elements until they eventually burn out and collapse, sometimes with a great explosion known as a "nova" or "supernova" which throws huge quantities of matter into space.

Gold is one of the heaviest elements, as demonstrated by the fact that it sinks to the bottom of the miner's pan. Scientists believe gold is formed in the nuclear cauldron of red stars, a phase which occurs towards the end of the stellar life span.

If all this is true, Mr. Graham's real carrot flakes have had quite an odyssey, their source gold having been hurled into space at fantastic speed many billions of years ago upon the death of an ancient star and from there travelling for countless light years across the universe to be captured by the gravity of our own sun, where the gold was finally condensed from a spinning cloud of dust and debris into the mass of planet earth.

Most of the heavy gold sank into the earth's core when the planet formed, but later some was lifted towards the surface by a convection current of molten magma. The current rose to press against the base of the young Appalachians, which were miles higher than they are today. Under tremendous pressure, the gold was injected along with streams of quartz into the myriad fissures and cracks produced by a long, narrow rift which underlay what is now the eastern edge of the mountains. While cracks up to twenty feet wide were filled by the intruding quartz, most were much smaller and the quartz could fill fissures narrower than a razor blade.

Over hundreds of millions of years, as the original towering peaks wore away, the removal of their tremendous weight allowed what had once been the core of the great mountains to rise. The gold-bearing strata along the great rift below was thrust skyward and itself eventually came to occupy a surface band running from southwest to northeast along the side of the by-then ancient range.

As the process of erosion continued over the aeons which followed, our tiny flakes of gold were finally released to begin a new journey down what became Mr. Richardson's creek. Somehow they had escaped a century and a half of mining operations to find a new home in the little glass bottle we bought for a nominal extra fee.

While we maneuvered the small flakes into the bottle, the goldminer proprietor of the panning place was addressing the question of whether or not the Indians mined gold in north Georgia. While the evidence is thin, many people nonetheless believe they did.

The proprietor told a story about a man who once used a metal detector to search an old farm further up Dukes Creek. The man came across what appeared to be three stone markers placed to form a triangle. After getting a strong reading on the long side of the triangle, he unearthed a clay pot which appeared to be of Indian origin. When the pot was lifted, it broke and a large quantity of gold spilled out.

The gold was split three ways between the man, his partner, and the owner of the farm. There was so much that they used a sugar scoop to divide it. Each man got 66 pounds of the precious metal, so there were almost 200 pounds of gold in the old pot.

The man with the metal detector was from down in middle Georgia, and that was where our gold-miner proprietor said he saw the man's gold and heard his story. Our miner also said that gold from different creeks looks different, and that the gold he saw looked like Dukes Creek gold.

All in all, it was an interesting afternoon at the panning place. We came away with a souvenir and got to hear a tale of the sort which have colored the gold region from the first days of the great Georgia Gold Rush.

The real carrot gold sat for a while in its little bottle on the shelf over Mr. Graham's desk. But since he could not resist pulling the tiny cork out of the bottle, its confinement was only temporary and the immutable metal was soon on its way once again.

And it should be noted that however grand the enlightened musings of modern scientists may be, Mr. Graham does not much believe gold comes from red stars. He thinks it probably comes from blue stars, since blue is his favorite color and blue stars are the best.

STORY OF THE "BEN FIELDS"

In the 1960s, some of us kids from Helen would often go to a place known as the "Ben Fields", a remote area which lay over on Dukes Creek. To get there we had to get around Hamby Mountain. Several rough roads ran through the high gaps at either end of the peak, crossing the old Hamby Mining Ditch high on the mountain's backside and from there descending past a moldering log cabin and the workings of long departed miners to reach the creek. The ancient fields had been abandoned for years, leaving the overgrown clearings as the domain of wild turkeys, deer, and an occasional bear.

The Ben Fields were special for us. Way back in the woods, they made a pretty spot where we had the freedom that young people need sometimes, with no one around to worry about what we did or didn't do. Like other deserted sites, the fields were imbued with a sense of mystery, leaving one to wonder what it must have been like years ago on a place where nobody any longer cared to live.

But this wasn't just any abandoned location, for it also occupied a region of its own in the collective local consciousness. The name was old and therefore sacred in nature, and we all knew instinctively that people who'd once lived there were ancestors to people we knew then. And since everybody local had always known of the Ben Fields, it was a thread in the tapestry which defined the community, and anyone who didn't know about them was probably from somewhere else.

Although we could peer a few generations into the past, it's not surprising that the human history of the Ben Fields is much older. The site was occupied long before the white man came, for archaeologists working there in the 1930s found Indian artifacts dating back many centuries across several distinct Indian cultures.[1] Also among the items recovered was a set of glass beads of European manufacture, a relic from the period which coincides with Cherokee occupation of the area.[2] Like the Helen valley and all of the Chattahoochee headwaters east of the Blue Ridge and the Chestatee River, the Dukes Creek area was obtained from the Cherokees by treaty in 1819 and distributed to Georgia citizens in the 1820 Land Lottery.

The Ben Fields stretched across several of the old Land Lots, but the first white man to own the homesite for which they were named was John Butt Sr. It was on Land Lot 68, which Butt bought in 1822 — duly noted on the deed as "the 46th year of independence of the United States of America" — for $300 from the man who had won it in the land lottery. Butt lived there until about 1833, at which time he joined a number of other local pioneers in migrating over the Blue Ridge to Union County, settling near what soon became the town of Blairsville.

Although Georgia was taking over, Union County was still in the Cherokee Nation when Butt arrived. For several years after his move, the unwelcome white intruders lived among the dispirited Cherokees. In 1836, three years after Butt and other Georgians had begun streaming in, Indian agents were sent to Union County to survey the value of Cherokee property in preparation for the Removal. A note they made shows the arrival of Butt and other Georgians had not been a pleasant occasion for the Indians. One of the natives encountered by the agents was a "fullblood old woman" named "Eutaeh", who told them she had been driven away three years before from her place where John Butt Sr. now lived, at the Coosa bend of the Nottley River.[3]

Before Butt left Dukes Creek, his property had attracted considerable interest from gold miners. In the spring of 1832, he leased a part of his property to a Tennessee man who was to mine for vein gold and give Butt 1/6 of any proceeds thus obtained. A few months later, in September of 1832, Butt struck a deal to sell his entire Land Lot outright to William Hume, a doctor from Charleston, South Carolina. In a story familiar along the Georgia gold belt, Hume did not pay cash, but instead gave Butt a note in the gold-fevered amount of $20,000. The ending of the story also has a familiar ring, for Hume never paid the note. Butt wound up keeping his Land Lot 68 property until his death in the mid 1850's, at which time it was sold at public auction.

The high bidder was Benjamin Wilson Allison, who paid $750 for the 250 acre lot. Family records say Benjamin was born in North Carolina in 1822 and was brought to Georgia two years later when his family followed in the footsteps of his grandfather, a pioneer who was already living on Dukes Creek. Ben W. was an active and accomplished man, for in addition to his work as a farmer and a goldminer, he was licensed to preach in the Methodist Church in 1850 and later served on the County Board of Education, as Judge of the Inferior Court, and as Justice of the Peace. Family records say Benjamin left Dukes Creek in his later years, going in 1888 to live with his son a few miles away near the town of Cleveland.[4]

Ben lived to be 82 years old. Upon his death in 1905, he was buried in the old Lawrence Cemetery, which lies across Dukes Creek from his old homeplace. For his long residence on the spot and his prominence in the community, Benjamin's name was permanently assigned to "The Ben Allison

Fields" or "The Ben Allison Place", with "Ben Fields" evolving as a shorthand way of referring to them.

In the 1890's, State geologists came by on several occasions to survey the holdings of a mining company which then controlled the Ben Allison Place and much of the surrounding area along Dukes Creek. They found thirty cottages in the vicinity, twelve of them frame buildings (presumably the rest were log cabins), and all "suitable for the occupancy of miners".[5] The geologists also found a fully equipped blacksmith shop and an unoccupied commissary building; these and the thirty dwellings were accessible via wagon roads. One report estimated that something between 10 and 20 million dollars worth of gold (at today's prices of about $400 an ounce) had been taken from the Dukes Creek mines in the six decades since the Gold Rush began.[6]

One geologist also described the agricultural scene along the creek. In typical geologist fashion, he calculated the potential yield of the fields just as he did for the surrounding ore deposits:

> The cleared, cultivated lands consist of approximately 218 acres. Of this, about 50 acres is fine bottom land, and would, if properly cultivated, produce 1500 bushels of corn. The remaining 168 acres ought to produce 1000 bushels, making a total yield of 2500 bushels. The farms, however, have been poorly cultivated, and the yields have never been what they should. Similar land, in the same vicinity, has sold for $40 per acre for farming purposes. Besides these cleared farms, there are approximately 300 acres of excellent farming lands which have never been cleared of their original forests. The farms. . . contain many convenient springs of freestone water, for drinking purposes.[7]

Some time after large-scale gold mining activity ceased around the Ben Fields, Henry Newton Abernathy rented the old place from its latest owner and settled in to raise a family there. Henry was the third child of Joel Abernathy, a pioneer who had lived in the Helen valley where he farmed lands owned by his friend and fellow Civil War veteran Sam Conley. Henry Newton and wife Emma Louise had also lived in the Helen valley when they first married, occupying a small house behind the home of J.R. Dean at the upper end of the Helen valley. Mrs. Dean had been a New England school teacher before coming to the valley, and it was she who gave Henry Abernathy what education he received as a boy.

One of the roads we used to reach the Ben Fields ran past the home of George and Laura Cannon, who lived beside the Orbit manufacturing plant on the north end of Helen. Twenty five years after our youthful adventures at the Ben Fields, my father told me Laura had lived there as a little girl. We took a tape recorder and went to see her.

Born in 1909 as Laura Abernathy, she was Henry Newton's daughter. Although we talked to her in the early 1990's, Laura's earliest memories were of a time not much different than that of the first settlers. The childhood tales relayed by her father and other elders were first-hand accounts of area life dating back to about the time of the Civil War, so she knew the stories behind many of the things she saw as a child.

AT HOME IN THE BEN FIELDS. Henry Newton and Emma Louise Abernathy standing in front of the old Ben Allison House. The house was covered with weatherboarding, but some of the logs used in its construction show on the front porch. Henry spent much of his life in the Helen valley, where as a boy he helped his father Joel Abernathy work the fields on the Conley Place. Henry and Emma also lived beside J.R. Dean when they first married.

And many of Laura's first memories were of things that dated back to the days of the first settlers and early miners. She described log cabins, her one-room school, miner's shacks, and the old "Parks Store" where the miners traded, all of which have so thoroughly vanished as to be virtually unde-tectable by anyone short of an archaeologist. The old Ben Allison home in which Laura lived had started out as a log cabin, but by the time she came along, it had seen such extensive weatherboarding and modifications with sawn boards that the old logs could hardly be seen. The house eventually burned and the site was later bulldozed to remove the remaining rock walls and walkways built by the old-timers.

Laura got more education than her father, attending three local schools before she was done. The first was less than two miles from her house, a one-room affair located on GA Alternate 75 (known locally as the Loudsville Road) where Dukes Creek Baptist Church is today. Laura also went to the new school in Helen and later attended the more distant "Nacoochee Institute", a Presbyterian enterprise once located where the valleys of Nacoochee and Sautee meet.

Since Laura brought a feminine perspective to things, some of her comments contrast with those made by the geologists who had visited her homeplace a few years before she was born. Where they found "habitations suitable for occupation by miners", Laura saw "old shacky houses", a clear indication that the rough abodes of miners were often not suitable for women and other more civilized people. And as for the "convenience" of the "free-stone springs", Laura disagreed, for she had to tote water home in a heavy wooden bucket and would have preferred a well closer to the house.

Laura was in her early 80's when we talked to her. She had lived at her Helen home for a long time, having settled in many years before with husband George Cannon to raise a family there. She was a wonderful person, witty and insightful and very much possessed of "a clear mind" as she would have put it. Laura had a great awareness of the things around her and could remember events in vivid detail. With an easy laugh, she was always able to find the humor in a situation.

Laura talked in what might now be considered the old way. Some might make fun of accents, but everybody has one. Laura's was a direct legacy of pioneers, something to be proud of, and a voice taken up long before modern mobility and mass communications began blurring things in the mountains. Within the severe limits imposed by the written word, Laura's recollections are recounted here just as she spoke them.

When the sawmill town of Helen was established in 1913, Laura was four years old. Along with the new town came the Gainesville and Northwestern Railroad, which crossed the Dukes Creek valley about three miles below the Ben Fields on its way to the Helen train station. If the wind was very still, the distance along the creek was short enough for the train whistle to be heard at the Ben Fields. This was a new sound in the valley, and in its travels, the noise covered more than miles.

The shiny rails of the G.& N.W.R.R. were the first major intrusion of an industrialized and rapidly modernizing civilization. As the shrieks of the train's steaming whistle rolled up the Dukes Creek valley, they passed acres of virgin forest and dozens of abandoned mines before reaching the ear of a little girl who played among stacks of hay and fodder on a farm where her family still made soap and cider, and the sound left her very curious about what she always called "the little train".

Today the Ben Fields are owned by the State, which manages them as part of the large "Dukes Creek Woods" natural area. Part of the old fields have been cleared again and can be reached via a paved road which now runs down the Dukes Creek valley from Alternate Highway 75. This road also gives a good view of many old placer mines further down the creek. Since the area is sometimes open to hunting and conservation is a priority, access is restricted. Contact the office at Dukes Creek Woods for more information.

Here, then, are the recollections of Laura Abernathy Cannon. They provide a last look in the rear view mirror to see an early lifestyle which has all but faded from living memory, and the final brush strokes in this sketch of life as it was during Helen's pioneer century.

THE RECOLLECTIONS OF LAURA CANNON

Our talks with Laura were recorded at her home beside Orbit Manufacturing Company at the upper end of the Helen valley. Her house was built on the same spot as that of J.R. Dean after the old Dean home burned in the early 1900s. The homesite overlooked the upper ford on the Unicoi Road, which crossed the river in front of today's Orbit facility. The narrative picks up as Laura is talking about her early years when she lived at the Ben Fields:

Laura: . . . we had a blacksmith shop. . . decorated Christmas trees with pop-corn. . . tied little gifts on the trees. . .

Q: Were the Ben Fields down there in the area that'd been gold mined?

Laura: Oh yeah.

Q: Was mining going on while you lived there?

Laura: The old pounding mill was straight across the river [Dukes Creek] from the old house, but it was not in operation then. It'd been out of order for some time.

Q: Were you up above Hamby Ford [a crossing on Dukes Creek]?

Laura: Yeah, we were further up than that. That did include down in there, all those flat places, those flatlands were in cultivation. They called 'em Middle Bottoms, and Lower Bottoms, it was all bottom land. It went on down, a pretty good little piece down the river. It joined the other old gold-mine, you take a left at what we call the "Whitehorses" [White Horse Gap on Hamby Mountain], after you go up to the top of the hill like going through by Greears, you know you take a left, where the big old deep goldmine is [big shaft at the Reynolds Vein], you take a left and go down to what they call the Carl White Fields. The Ben Allison Fields joined that, there was a stamp mill down there too.

Q: Was that stamp mill at the Reynolds Vein?

Laura: That was the place, over in there. The mines had already been cut when we lived over there. It was just a big farm. Right along the fence, the stock, they just roamed the woods, you know there was no stock law. Everybody turned their pigs and their cows out, and what they wanted to grow stuff, they had to fence it. Everything was fenced in. It was just a big place. And it had four pretty livable houses. One big house, one we called the big house, it was the same house Ben Allison built. It had a big living room with a attic up over it, then a big room in the back, kinda like a barracks, you had several beds in the back of the house. It was made of logs and sawed boards, mostly sawed boards. We had a big cellar under the house. And then the kitchen and dining room, it was a separated building, you had a little breezeway that you went through and it was just one room. And they heated it with a big fireplace. My father rented the whole plantation for $85 a year. He had subrenters. He furnished all the stock and they would tend the fields. They would take two rows of the corn and he got one row. And they would pile it all up in a big shed in the fall of the year, had a big log crib, and then we'd have corn shuckings. Everybody'd go to the corn shucking. The ladies, they'd cook and make everything, they'd have big pots of turnips, n' beans, potatoes, a little stuff like that, along with dishpans of pie. There was nothing to do, and the boys and girls would all go to these corn shuckings, and if they found a red ear of corn, well, all the boys got to kiss their sweetheart [laughs]. And they'd get the corn shucked that way.

Q: Were there many red ears of corn? Must not have been too many. . .

Laura: Every once in a while you'd find a red ear of corn in a big pile of corn. So, that was a lot of fun. On Saturday, there was a big orchard on the place. On Saturday afternoons, the day everybody'd pick apples up. Some'd wash em, some'd be a cuttin' the rotten spots out. They had just a square box, it was about like this [holds up hands to show size, about two feet square]. They had a wooden mall. And they'd put the apples in there. One would mash the apples up, they'd beat the apples with the wooden mall. Then when they got the apples all mashed, they had a hole in a big walnut tree and they had a pole that was tipped to fit that hole. And they would stick it in that hole. The box was real tight, they'd put a lid real tight, then they'd put another board under that pole. Everybody would get on that pole, cider would just come oozing out, we'd make big jugs of cider and put it in the spring. On Sunday afternoon then everybody would gather and have cider. The children would all gather and climb trees and play.

Q: How about food you needed to keep cool?

Laura: Kept it in the spring or in the cellar.

Q: I've heard that, in the old days, people would kill a hog and divide it all up. . .

Laura: Oh yeah. Everybody would get a big bucket. They used a lot of buckets then, there's not containers like they are now. And cut up, and all the neighbors'd get meat. [Later, the neighbors would return the favor, ensuring everyone a steady supply of fresh meat.] And they had bees, and had bee robbin's, and neighbors'd come in and everbody'd take back a supply a' honey. The bees were from beehives, they had their place where they had their beehives.

Q: Did you ever hear of bears getting any of those hives?

Laura: No, we didn't even hear of bears then. [Although they've since made a remarkable comeback, bears were nearly extinct in north Georgia in the early 1900s]

Q: Not too many years ago, they'd let school out in September and have fodder pullin's. . .

Laura: Yeah, and pick peas. Everybody had peas then. Fodder [the leaves on the corn stalk] was gathered because they had to feed their stock that. They'd take 'em and tie 'em in a knot and let 'em dry out and hang it on a stalk, they'd have both hands full and tie it on a stalk and it would dry. Later, after they'd dried out about two days, in the evening, when it was cool in the evening, and it was getting damp, they'd take four of those handfuls and put it together and make a bundle of fodder. Everybody had to gather fodder, and they'd take it to a pole in the ground, and stacked it around. One'd get up and mash the fodder down as you threw it up. You threw the bundles of fodder up to 'em and you'd make a big stack and you'd taper it to where it'd shed the rain.

Q: Back in that time there were a lot of chestnut trees. Did the stock eat the chestnuts?

Laura: The wild hogs did.

Q: Do you think you could raise hogs out in the woods now?

Laura: No. . . no I don't think so. The chestnuts are gone. Hogs will eat acorns.

Q: What was the Ben Allison place like?

Laura: We lived in the big house. Ben Allison was the one who built the place. They'd been a lot of work put in the place. They had a rock wall with a flower garden, with a walk going up to the other house all made out of stone. They had a lot of pretty shrubbery on each side of the walk. The lower part, they had another rock wall. You had to walk a long ways to the spring even though they could have dug a well. I think it was an old log house, you could look down from upstairs and see the logs inside it, with weatherboarding lumber on the outside of it. It was an old house when we lived there.

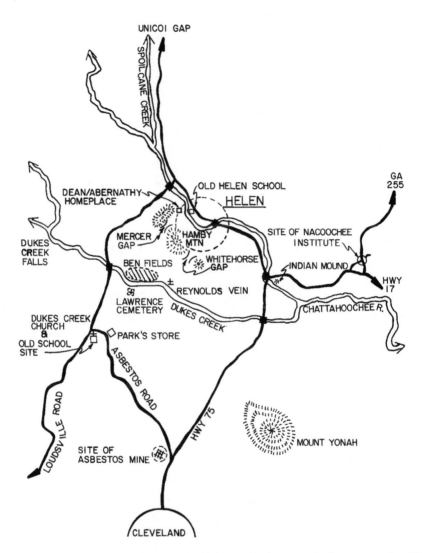

LAURA'S WORLD. Several roads led through the woods between the Ben Fields and the Helen valley, including ones through Mercer and Whitehorse Gaps at either end of Hamby Mountain. Laura attended three schools: first a small, one room affair located where the Dukes Creek Baptist Church is today and later the old Helen school and the Nacoochee Institute.

Q: Was a lot of that area over there at Dukes Creek cleared land?

Laura: After it was mined, all those fields, those bottoms [were cleared]. . .
now the river has cut in and took a lot in front of the old homeplace. There
was gold mines all around, but everything was in cultivation. They's no trees,
it was most all open land, they was a meadow that grew wild grass, and mowing that, cut the meadow and stack the hay. We had a mowing blade, cut it
with a mowing blade. The meadow was as big as out here where Orbit is [a
sizeable bottom area]. The McConnell place went with it too. There was a
big house up there, a big log house.

Q: Was the place self sufficient, providing all its needs for its people, the food
and medicines. . .

Laura: No, it didn't provide medicine. We used the doctors here in Helen
[after 1913]. They went with their little pill bags wherever they was sickness.
And they'd walk in to that old place. [Although Laura's family was using
"real" doctors, folk medicine was also practiced in the mountains, the
providers of which were sometimes referred to as "conjurers".]

Q: You had yellowroot. . . [used to make a medicinal tea]

Laura: Oh, yeah, had yellowroot, and they was a tea, a tree that grew along
Dukes Creek that made wonderful tea.

Q: Was it Sweet Birch?

Laura: No, it was not Birch.

Q: Sassafrass?

Laura: No, it was not sassafrass either. It tasted a lot like the tea we buy.

Q: Were there stores over there [on Dukes Creek]?

Laura: Yep, the Parks Store, [husband] George's Grandfather had the Parks
Store. They was a lot of houses over in there. And the Parks Store was located not far from the garbage dump [White County Landfill] and there used to
be a big old log house where you turned there to go to the dump [Parks Store
was a little over a mile from the Ben Fields]. They sold everything to the miners, it was built for the gold miners. You know, then it was gold rush [when
the store was in operation] and everybody would throw up these little shacky
houses and they was just a lot of them. . .

REYNOLDS VEIN. This is the largest vein working in the Helen area. Ore from the vein was crushed with a 20-stamp mill powered with water from the Hamby Ditch. Although this hole at the surface is small, the shaft opens into a large room below which disappears below the water line. The road Laura traveled from the Ben Fields to Whitehorse Gap and the Helen school passed near this abandoned mine.

Q: Who started the Lawrence Cemetery over there across from the Ben Fields. . ?

Laura: It was a Methodist denomination, but I don't know who started it. They's a lot of Allisons in that cemetery, so I would think maybe it was started by the Allisons. There may have been a little church near that cemetery, but there was no church in the time I remember. . . . At the time gold mining was prosperous, which was before I remember, they would people move in and work the gold mining and they'd just build little shacky houses. I've heard my father talk and he'd say some fellow used to live in a little shacky house there, just all over that old place.

Q: Was your father Henry Abernathy?

Laura: He went by the name of Henry Newton Abernathy. He worked in the gold mines when he was a young man, that was all they was to do. He got his education by working, what little bit of education he had, he never went to school a day in his life. Now my mother did, they had a little one room schoolhouse and one book, Webster's Speller I think they called it. She went, she had a little education, but he didn't. But they was a Mrs. Dean that was a school teacher and she had a son the same age as my father and she taught him. Because the schools were so far between, you'd have to go for miles to get to a school house. The little school I went to was over there at Dukes Creek and was just one little room and a big old stove in the middle of the room with a sandbox around it on account of fire, to keep fire down, had just old benches, you just sit on an old bench and didn't have a desk.

Q: The school you went to, where was it?

Laura: They's a little church there now. It's up on the road, it's on that road that goes through by Loudsville [Alternate GA 75]. The church is right on the bank.

Q: That church where the Asbestos road comes in? [The Dukes Creek Baptist Church, located where the Asbestos Road ends at the Loudsville Road/GA 75 Alternate]

Laura: Yeah, right there. Right where you turn off. And then that's the old John Ledford house. But they was a big log house on the main highway where you turn off, too. . . it's all rotted down. The little schoolhouse was just a little further back than the church. And we had a spring down there not far from where the highway goes through. We'd get a bucket of water and they'd be a dipper and everybody would drink the same water out of the same dipper.

Q: You had to go a pretty good little ways to get to the school.

Laura: Yeah, and we had to walk. Started in July and we'd go through July and August and then we would stop. After everything was gathered in, the hard part of winter we went to school and then when it was time to go to farming again, we'd have to drop out of school. Altogether, we only went seven months. Everybody had to work.

Q: The school was on the Asbestos Road — where is Asbestos?

Laura: They's an asbestos mine down in there. I never have figured out exactly where it was, but I have a friend who's 91, and I asked her where the asbestos mine was, I'd heard of it all my life, but I never [knew], and she says that, you know where that little church is, where you come out that Asbestos Road into [Highway] 75? That little Holiness Church? Well she says it's just right up above it a little bit, kindly in behind it, right in there. [One of several in the county, this asbestos mine was located near the lower end of the Asbestos Road where it joins Highway 75 near White County Park.]

Q: Was there a school in Helen then?

Laura: They was a little school there where the motel is [The Chattahoochee Motel at the north end of town]. They had an old pump, you pumped water [drinking water from a well]. A lot of times we would finish out our nine months school, we'd walk across the mountain when our little school in Dukes Creek would give out, well we would walk and finish out some of our school in Helen. We'd walk across the mountain [Hamby Mountain] to school. We mostly came through by Greears [Whitehorse Gap], it's a little shorter that way. They would be a hundred people, it was a pretty big school house, I don't think it was cut up in rooms, just a big schoolhouse, and we would have Sunday school and they'd be a hundred people come to Sunday school. They's nothing to do for nobody and everbody went. They didn't have cars to go other places. And everbody went to church. The school was where the Chattahoochee Motel is, right about where the motel buildings go along the right [beside Hamby Street], the school was pretty close there.

Q: What about the Nacoochee School?

Laura: When I went to Nacoochee School, we went in an open Ford car, you know just a top pulled up, a cloth top. And we'd go out the coldest mornings, bundle up and go to school. I lived here in Helen then. I was about thirteen years old. The [Nacoochee] Institute was still an Institute then, Raburn Gap had not been built. The school had burned. The school had burned in April before that. And the school was where that store is down there [Sautee grocery store]. And the boys dormitory, they turned it into classrooms. I went to the boys dormitory, it burned, and shortly after that they moved it to Rabun Gap. [The Rabun Gap - Nacoochee School still operates in Dillard, GA]

Q: Did you make all your lard, stuff like that. . . ?

Laura: Soap. You took your ashes, you didn't burn anything but oak wood, went out and cut prime oak wood, and fire ye' fireplace, two fireplaces, and as you took ya' ashes out you put 'em in a hopper we called it. It was big boards, they split big boards this wide [holds up hands to show size of shakes] out of trees you know to cover the roof, do repairing of it. Always they was a stack of boards. You'd take those boards and you'd make a hopper and it'd be about from here to here and stand it up like that [at an angle] and pour your ashes in there. And it was at a certain time of the moon in March that you dripped your ashes down. And you had broomsage in the bottom of your hopper where the ashes wouldn't go through. You had a board that you'd made [into] a little trough that your hopper was sitting on. And you put a pan under that and you wet your ashes. And you'd start just dampening them and you'd just dampen them until they'd get real wet all around in your hopper, just put water in and let it soak in slowly til your ashes would all get wet and you'd go to gettin', oh, not more than two dipperfuls at the time, go very slowly with the water. And after while you'd see a little stream of real red dark lye start comin' out. Maybe you'd drip all day, I know my job was. . . ash-hopper, and I'd play around and when time come, I'd pour my water in the ash-hopper. And when the pan got full you poured it in a churn cause it was very strong, it's like Red-Devil lye. They's another time of the moon, shortly after that, it was in March you made your soap for the whole summer. You took that lye and you saved all the drippings, when you kill a pig, and have old stuff's no good to eat, you could still fry it out and make grease out of it. You'd take all the drippings that you'd saved through the summer and the winter and you'd put it in a washpot. I don't know the process, but when you got through with it you could just take and add that lye to it, and the grease would sometimes rise to the top, it wouldn't take all the grease, so mother would take and scoop all that grease off, and then she'd take and get a big long knife and she'd start cutting out soap, bars a' soap you know.

Q: That'd take the hide off though, wouldn't it?

Laura: It'd eat your hands, you'd be careful a' handlin' it.

Q: You remember leather britches, and dried apples. . .

Laura: . . . I still like to dry apples. They used to take cabbages, turn 'em upside down and throw dirt on top of 'em, and preserve 'em all winter long. They'd dig their potaters and they'd scoop out a big hole and line that hole with hay — straw — and pile their potaters all in there and put more straw on the top and they'd cover it with dirt and put a top over it to keep it from rainin' down in it. And they'd leave a hole and they'd take a burlap bag and stick in that hole. You could pull that hole out an' go get ya' potaters all winter. And your turnips, you did your turnips that way, fresh turnips. You didn't have nothing to buy but a little coffee and sugar.

Q: Did you have a mule or did you have an ox?

Laura: Oh, we had a mule. But I remember seein', they was one old man, he drove an ox wagon an he hauled lumber and he'd go by and the old oxen'd just be barely movin'. He had two big fine oxen. But one'd got his tail hung, and had a little tail about like that [holds up hands to show how short it was]. He'd wiggle that little tail. This old man'd ride along and he'd go to sleep on his lumber, and some little boys was sittin' on a fence and laughing one time as he rode by and he woke up and said, "What you laughing at?" They said "nothing!, nothing!" — they didn't want him to know they's laughing at his steer. And he says "I know what ya laughin' at — you're laughing at my steer's tail." They said "No sir — your steer don't have a tail!" [laughs].

Q: I've heard some old timers say they would never hit a wedge with another steel instrument. They had a wooden mallet, cause they couldn't afford to ruin the wedge. . .

Laura: . . . They called it a wooden mall. They'd get a big tree and saw off of it about this long [about three feet] and then they'd cut around it, and they'd work the handle down, it'd be one piece of wood. That's what they'd beat the cider with, you know.

Q: That must have been some hard wood — what did you use, oak?

Laura: Hickory. Oh, you could make out of oak, but everbody preferred hickory.

Q: Did you ever dry any corn?

Laura: No, I've heard people talk about it. . . . Now pumpkin, that was the thing everybody dried. They'd cut their pumkin in little thin strips and they's take a pole and string their pumpkin up on that pole and fasten the pole somewhere where the sun would shine on it and they'd dry their pumpkin. They'd put it in a bag after it dried and store it for winter. Another thing they had, everbody had peas and they just had, oh, bushels of peas. They would take the peas, they'd pick ya' peas and when they was real dry they'd put 'em on a sheet, some kind of a heavy sheet. The fertilizer was always in heavy bags, cloth bags, real heavy bags. . .

Q: What kind of fertilizer was it?

Laura: Well, guyana, they used a lot of guyana. . . . Well, anyway, they'd take those big old heavy bags and sew 'em together and make a big sheet. And they'd lay the peas out on there and then they'd get 'em a nice big stick, everbody'd get around and beat them peas. And they tried to do it on a windy day. Those hulls would all beat off and then they'd take the peas and lift 'em up like that [swings arms to show how everyone would hold on to the sides of the sheet and throw them up in the air], and the wind'd just blow all those hulls out [leaving the heavier peas to fall back on the sheet]. Then you'd put your peas in a bag. The cloth'd be about like a bed sheet, maybe a little bigger than a bed sheet. That's the way you hulled ya' peas. And they was no such a thing as bean beetles, you could just throw down bean seed or anything and, oh, they'd just grow and you'd have just have more beans and things than you could use. And now you have to fight the beetles for everthing.

Q: Oh, the Japanese Beetles?

Laura: Uh-huh, they was no such a thing as a Mexican Beetle or a Japanese Beetle. I remember when the Mexican Beetle first hit, it just eat everbody's beans up, they didn't know how to fight it.

Q: Have you been back to the Ben Fields lately?

Laura: It's all changed. You know, I didn't go over in there for years and when I went I hardly knew the old place. Marvin and I went over there, my brother. . . When I left over there they was fruit trees and everything was open. . . everthing was so growed up. But I had been back since he had and I told him we could drive all the way down to the lower end of the field then and I told him, I said now here's where the old house is. He said no, it's on further down. He's drivin' and he went on down and we got out at the old spring. . . We carried water from a spring. I said see, there's the old spring, we're way down below the old house place. Oh, no, no, you're mistaken, it's on down futher. And I said well let's just walk down through here. See, you couldn't see no spring, just a wet place. And they used to be a big rock, well it's still there at the bend of the river [Dukes Creek]. . . My mother caught a fish this long [holds up hands to show big fish] off the old rock and she start-ed screaming and everybody thought she'd fell off the rock in the river. And I said see, here's the old big rock. And I said here's the spring branch. And he said, no you're mistaken. But just to satisfy you, we'll walk back up this way. So we walked back up there. Now, Smithgall has pushed all the old walls down, [but] at that time they were still there. And we got back up there. . , it's all grown up you know, and we had to fight through honeysuckle and briars and everthing. He said, now I wouldn't go through this only just to show you you're mistaken. So we got up there and there was the old rock wall and he said, well I'm ashamed to have to admit it but you're right. You just wouldn't know the old place now like it was then. [part of the old fields have since been re-cleared]

Q: Why did it grow up like that?

Laura: Well, after the Whites bought it, well we moved — my father had bought a place up on the road, just a small farm because the family was a decreasin' and he knew he couldn't handle the big farm. He had invested in some property up there that had a little house on it. So, the Whites, they changed everthing around, and he didn't like the setup, so he moved. The old house stood down there but just first one and another lived in it, and they bushwacked it around, and nobody took any interest in cultivating any of the land, and finally the old house burned.

YOUNG LAURA AND FRIENDS. Looking like a group not to be messed with, from left to right the girls are Mae Stephens, Flonie Allison, and Laura Abernathy.

Q: What about the rest of the property down there? Did those people move out?

Laura: Well all the old houses just fell in. The Parks Store, now when we lived over there, they was just a family lived in the Parks store.

Q: That was over near the landfill?

Laura: Yeah, it was in that direction. You went out the road like you were goin' to the dump. I can't place just exactly where it's at. Mr. Slaton had Parks store and he bought everbody's gold, people would gold mine all week and then take it to the Parks Store.

Q: Did they sell a lot of gold in there?

Laura: Oh, yeah. And he would buy their gold and furnish 'em groceries, what they had to have.

Q: Did you ever ride the train? [The Gainesville and Northwestern Railroad came to Helen and Robertstown in 1913]

Laura: My mother would always go onest a year to Cleveland and pay our rent. And we would ride from Helen to Cleveland and that's as far as I ever did ride on it.

Q: When you rode the train to Cleveland, was that a big adventure?
Laura: Oh yeah. We'd walk all the way from Dukes Creek to catch the little train to ride to Cleveland, but you could have almost walked to Cleveland [laughs]. We got on at Robertstown. I always looked forward to goin' and payin' the rent. We paid it to McMillian.

Q: Do you remember how much the rent was?

Laura: $85 a year for that whole plantation [laughs]. It was a big scope of bottom land. But now it's all grown up in trees. We needed to a' made a lot of pictures back then, but we didn't.

Q: Did people have cameras back then. . . ?

Laura: Well, people didn't have cameras, but a lot of times, I remember when I was a child, they would a man come through makin' pictures you know and you'd have your picture made. They'd walk, everbody'd walk, it was nothing for somebody to come through a sellin' something, or a makin' pictures, and spend the night with ya'. They was no places to stay. They would drive wagons all the way from Hiawassee to Gainesville and pick up produce. And our place was about middle ways and forever somebody from Hiawassee was spending the night. They would usually come through the Loudsville Road, and the Asbestos [road], a lot of them would. The road over here [describing the old Unicoi Road through Helen, she points to the upper ford in front of her house], this road went right along the edge of the river, a little narrow road, and you went just right along the river, and down here right in front of the house was where you forded the river, and you went down and you came out at that eating place and you forded the river. If you was in a wagon, buggy or anything, you forded the river down here in front of my house and kindly went down the river and out at that place of Hammerson's [The Hofbrau House]. The main street, that was just a little dirt road, that road was there. They was a bridge [where the Main Street bridge is today] after so long a time, but they was a long time they forded the river again. They had a swinging footlog that the people could walk across, but the wagons had to ford the river, again. So, a lot of times they'd go this Loudsville Road on account of avoidin' so much crossin' the river.

Q: How often did you take a bath back in those days?

Laura: Well, you had a tub. You had to have an old washtub. And about the washin', I bet you wouldn't know about the way people washed back then. They had what they called a wash place. And they had a good sturdy bainch [bench], maybe it wauldn't quite as tall as this table and they had a piece of wood you know, made into a paddle, like you'd paddle a boat with or something, or bat a ball with, you'd put your clothes, you'd build a fire around the washpot and you'd get your water all hot and then pull in your tubs and you'd rub yore clothes you know, you'd rub 'em and rub the dirty spots, and then you'd take 'em and put 'em on this bainch and just beat, you'd just beat 'em good with that paddle and you'd get all the dirt out you could and then you'd put 'em in the wash pot. And you'd have a big rolling fire you know, and you'd put this high powered soap in that you'd made. And you'd just let 'em boil and you'd push 'em down with your stick and let 'em boil good and then you'd take ya stick and you'd lift a few out and you' put 'em on that bainch and you'd beat 'em again. And then you'd put 'em in water and you'd wash 'em out of that water and you had to wash them through and you'd rub the dirty spots again and beat some more, you just beat ya' clothes to death, ever' button on your clothes'd be bursted. And then you'd rinse 'em, they had to go through at least three waters. But they'd be as white as snow. You wore your clothes until they needed washin'. And then the scrub board came by, you know, you could rub 'em on the scrub board. That was somethin' new, that scrub board, everbody had to have a scrub board. And now that's a thing of the past. And you'd wash all day, it'd take ya' all day. You'd let your dirty clothes accumulate, you know, and then you'd just tie a big sheet full of clothes and you'd wash your white clothes, and wash your colored clothes, I mean a tub. And, oh, you'd have your clothesline all the way across your yard, hanging with clothes. But you didn't wash as much as you do now.

Q: You'd wear a shirt or dress or something two or three times...?

Laura: You'd wear it until it got good and dirty, however long it was, you didn't time it. But when it got dirty, you'd find something clean, put that in the dirty clothes. . . .

Q: Was it harder to wash in the wintertime?

Laura: Oh yeah, it was hard to wash in the wintertime. And people talk about going back to the old timey ways, now, back then we didn't know no better and it was fine. Everybody enjoyed life. But now I wouldn't want to go back to the old days. I don't see how older people made it, you know, having to wash outside, and the way they'd have to do, have to live. . . it takes a lot of strength to have to do that.

Q: What about you taking a bath? Did you use that same soap to take a bath?

Laura: No, we bought soap for our bath, because that would take the hide off ya'. But washin' your clothes, ya' made your soap for the year.

Q: What kind of soap was it you bought — do you remember a brand name?

Laura: Yeah, Octagon. Everybody used old Octagon soap. And it was not far from the homemade soap [laughs].

Q: When you took a bath, was it in a washtub?

Laura: Washtub, yeah. Heat up water and pour it in there. And us kids'd get in. . . And most of the adults would get 'em a pail of water, you know, and they had these little tubs, foot tubs we'd call 'em, or a large bucket or something, and they'd just a' sponge, you know, just take a washcloth and sponge theirself.

Q: How often did you take a bath?

Laura: ust whenever you needed to, once a week whether you needed to or not. . . [laughs]. Nobody knew any better, that's the way they grew up and that was just a way of life.

Q: Did people use deodorants and stuff like they do now?

Laura: No, . . no. . . it was a smelly crowd [laughs]. Everybody smelt the same though. Nobody noticed it because everybody was alike, it didn't make any difference [laughs]. . . you just stayed back in the woods. Maybe you wouldn't see anybody all week. . . . It was kind of a hard life. . . but everbody worked, that's one thing everybody done, from the time you was little you had your amount of work you was allotted to, as what you could do, and you just grew up and everbody knew how to pitch in and work, no problem.

Q: Do you think young folks have as much to do now days?

Laura: No, and they don't know how to do. They get bored and they don't have nothin' to do but get in to something. No, the young folks don't have any responsibilities, because it's easy for the parents to just to go ahead and do all the work and let 'em go free.

Q: Were there a lot of people around who lived to be a pretty good age?

Laura: No, they died young, because they was a lot of sickness. People died at a young age because they didn't have medical science to keep them alive. And it was not a long life. A lot of little children died, you know, at a young age. . .

Chapter 8: *Epilogue*

ANCIENT SIGNS, OLD VOICES

It's just a small "scope of bottomland", but a lot has happened in the shadow of Hamby Mountain since the white man arrived and began keeping records. The Helen valley has been an interesting intersection, one where the Chattahoochee River and the Unicoi Road met the Georgia gold belt. For a century after it was built along the route of an ancient trail, the Unicoi Road was the major thoroughfare, leading both the first settlers and a hundred- year-long parade of strangers along the headwaters of the Chattahoochee.

In the time well beyond the reach of living memory, babies were born and pioneer parents left behind as generations of now-forgotten families came and went. Slaves toiled for decades in fields and mines before exiting after the Civil War to face the daunting prospect of freedom with little to prepare them for the experience. Through years of ups and downs many gold miners lost money, but a lot of the loss was borne by Yankees and even foreigners, leaving the area with its share of local winners in the great Gold Rush.

By 1900, with no descendants living in the Helen valley, memories of the original settlers were growing dim. Some of their names still attached to the small creeks in the valley, but the England Cemetery was abandoned. Where Englands, Bells, Conleys and Pitners had once lived, other families had come to occupy their old houses and toil in the fields they had worked to clear. Also of pioneer stock, these newcomers lived in much the same way as their predecessors.

After a century of relative isolation, a railroad came to the Chattahoochee headwaters in 1913. The railroad and the great sawmill which spawned it brought many outsiders and the trappings of a different and much more modern world. As Helen became a town with phones and electricity, the things of pioneers were swept away so thoroughly that the only remaining identifiable items known to have been touched by their hands were the small gravestones in the England Cemetery.

Times were changing, but all things old did not immediately disappear. For many years, wagons rolled alongside the tracks of the Gainesville and Northwestern Railroad and passed automobiles on narrow, rough roads. In the 1930s there was still a blacksmith in the Helen valley and a wheelwright nearby. The old Unicoi Road was not replaced with the modern paved route until the eve of World War Two, and it took years beyond that for many in the mountains to get electricity and indoor plumbing.

The boom produced by the timber industry lasted only about two decades. The lumber men came for the prime trees in the virgin woods. Once the area was cut over, the days of the great mill were done. As Charles White wrote in an essay for the White County Historical Society:

> Morse Brothers sawed their last board at Helen on May 5, 1931. At two o'clock in the afternoon of that day, the head of steam in the boilers was released by the whistle, which blew for thirty minutes, and when its echoes died out among the mountains, an era in White County died too.

When the tracks of the G.& N.W.R.R. were pulled up a few years later, the train was gone as well. Helen had become a town and been a wild and busy place for a while, but as things slowed down many of the newcomers moved on for other opportunities. It was a situation not entirely unlike the Gold Rush of a century before. Many came to the area and some stayed, but most left when things got tight.

As things quieted down, old voices remained. The descendants of original settlers who had been there before the mill came were still around when it left. The Unicoi Road, the Gold Rush, the impress of the Twentieth Century; all of these things had an impact, but slowly evolving accents brought by long departed pioneers continued to color the speech most often heard and would do so for a while.

In all the years since the train came, people have been talking about how much things have changed. Since perspectives start with one's own experience, most use their youth as the first benchmark to establish "how it used to be". For me, this period was thirty years after the mill and the train left. In the Helen/Nacoochee area of the 1960s, most families still had more cousins than even they could count, except at the annual reunion. As they did in other isolated rural areas, everyone looked and waved when passing on the road, whether they knew each other or not.

Even though pioneer habitations were just about gone, other signs of the past tended to last for a while. Here and there stretches of the old Unicoi Road remained, bypassed by modern road builders. Except for a covering of laurel, the excavations made a century and a half before by a long forgotten gold miner would be just about as he had left them. Although the buildings which once housed them had long since rotted down, large millstones painstakingly cut generations before survived to be imbedded in contemporary masonry or displayed as yard ornaments.

Although the mountains had always lagged behind in acquiring the trappings of the modern age, by the 1960s the gap had shrunk from decades to years. It was a situation which made for some interesting contrasts. As recounted in the noted *Foxfire* books and portrayed in Hollywood movies, in the Appalachians the "plain living" lifestyle of the pioneers lasted well into the 1900s. The railroad brought change to the Helen area a little sooner, but the older folks there had nonetheless grown up at a time when they made their own soap from ashes and fat. But in the advertising age, they were nearly as likely as anyone else to be using a dozen different newfangled soaps and cleansers which had by then been found necessary for the well-being of the populace.

Most homes had a television, but often the smokehouse still stood out back. There were contemporary "houses of beauty", but old women still dyed their hair with walnut hulls. When someone had a wart, they could go to a conjurer who'd rub it with an onion, which he'd then bury under a trash can. The wart usually went away. A man with emphysema might get a fancy electric machine from the hospital through which he could breathe and receive metered doses of oxygen and medicine. Such contraptions helped a lot, but even so, it would be no surprise if he also turned to an older remedy, heating dried jimson weed in a coffee can upon the stove and inhaling the fumes.

In the years since 1970, change has put its blurring foot on the accelerator pedal. The days of plain living have all but faded from living memory. White County's population has doubled in the last twenty-five years; some of today's older folks remember the 1970s as being the last time they could say they "knew just about everybody". Chain stores have come in to replace many locally owned businesses. Family reunions are on the decline and people don't volunteer the communal wave along the road like they used to.

Many express regret over some of the changes which have come to the Georgia hills. As modern times drop ever-larger pebbles in the pond, the ripples make old accents harder to hear. But in the local schools, young people still tend to talk with a unique mountain twang. As long as echoes of old voices endure, unbroken threads wind across the generations and something of the legacy of pioneers remains.

And even though the four-lanes and the electric flash of modern life tend to overrun old ways, sometimes things are just overshadowed, leaving the past with us in more ways than it might seem. Dr. Matthew Stephenson would probably be glad to know that there are many mountain churches which remain beyond "the cursed influence of European modern philosophy", places where evolution is not even a footnote to creation and eternal truths are unchanged at least from the Great Awakening of two centuries ago.

Although they may draw a crowd, waterfalls still dance across the "upheaved strata" and much "charming mountain scenery" enjoys a measure of protection in the National Forest. Change comes faster, but the mountains have always been special. Hopefully they will remain so. I hope also that this journey into the "dwelling place of the phantom things that were" proves useful in reading the ancient signs and hearing the old voices which can yet be found on the headwaters of the Chattahoochee.

ENDNOTES

NOTES FOR CHAPTER 2: FROM THE CHOTA TO THE CHATTAHOOCHEE

1. James Mooney, *Myths of the Cherokee, in Nineteenth Annual Report of the Bureau of American Ethnology, 1897-98. Part 1.* (Washington, DC: Government Printing Office, 1900; full text reissued Asheville, NC: Bright Mountain Books, 1992), p. 526.
2. Dr. Matthew F. Stephenson, *Geology and Mineralogy of Georgia* (Atlanta: Globe Publishing Company, 1871), pp. 104-5.
3. *Funk and Wagnalls New Encyclopedia,* 1986, s.v. "Sidney Lanier".
4. Dr. John Goff, *Placenames of Georgia: Essays of John Goff*, ed. by Francis Lee Utley and Marion R. Hemperly (Athens, GA: University of Georgia Press, 1975), pp. 335-39.
5. One such map is Eleazer Early's 1818 "Map of The State of Georgia". A copy is in the Ivan Allen Collection at the Georgia Tech library in Atlanta.
6. The 1820 survey of the 3rd District of Habersham County was done by A.C. McKinley. His original notes and a microfilm copy are on file at the Georgia Department of Archives and History in Atlanta.
7. Adiel Sherwood, *Gazetteer of Georgia,* 1837, pp. 140-2.
8. Mooney, *Myths,* p. 511.
9. Stephenson, *Geology and Mineralogy,* p. 73.

NOTES FOR CHAPTER 3: A PARADE OF STRANGERS

1. James Mooney, *Myths of the Cherokees,* p. 87.
2. The Dr. John Goff Collection consisting of original loose papers, printed items and graphics is stored at the Georgia Department of Archives and History. The section on the Unicoi Road is found in Box 11, File 27. The quote here is from an 1817 letter reproduced in the 9/17/39 edition of the *Chattahoochee Times.*
3. One reference to this early trail is in Chandler, *The Colonial Records of the State of Georgia,* vol. XXII, pt. 2 (Atlanta: Chas P. Byrd, 1913), p. 245.
4. Mooney, *Myths,* p. 542.
5. Interview with Laura Abernathy Cannon at her Helen home, August, 1992.
6. Goff, *Placenames of Georgia,* pp. 62-7.
7. Colonel George Chicken's journal of his travels through the Cherokee Nation in the winter of 1715-16 was reproduced in the *City of Charleston Year Book* of 1894. This excerpt is on p. 541. Copy of journal pages from Yearbook obtained courtesy of City of Charleston Historical Society.
8. From a handwritten account by a militiaman believed to be "Arthur Fairies" (signature is hard to read). Account is contained in *Draper Manuscripts, Thomas Sumpter Papers, Series VV,* vol. 3. The Draper Manuscripts are on microfilm at the Emory University Library in Atlanta.
9. Robert Eldridge Bouwman, *Traveler's Rest and the Tugaloo Crossroads* (Atlanta: State of Georgia Parks, Recreation, and Historic Sites, Historic Preservation Section, 1980), p. 91.
10. Lucius Q.C. Lamar, *Compilation of the Laws of the State of Georgia, 1810-1819,* "Act No. 489" (Augusta: T.S. Shannon, 1821), pp. 774-6.
11. Adiel Sherwood, *Gazetteer of Georgia,* 1837, p. 141.
12. Bouwman, *Traveler's Rest,* p. 93.
13. For extended descriptions of life along the old turnpikes, see Wilma Updike, *The French Broad* (Knoxville, University of Tennessee Press, 1966), chap. 9 and Bouwman, *Travelers Rest,* chap. 6.
14. E. Merton Coulter, *Auraria: The Story of a Georgia Gold Mining Town* (Athens, GA: University of Georgia Press, 1956), pp. 21-2, and Dr. Tom Lumsden, interview, Nacoochee Valley, August 1995.
15. Lumsden, interview.
16. Goff, *Placenames,* pp. 284-5.
17. Information on the final days of the Unicoi Road and Smith Crumley provided by Col.

Comer Vandiver in an interview at the site of the last gate on the Unicoi Road, August, 1994.
18. "Spoilt Cane Special", *The Cleveland Courier,* 22 June, 1917, p. 1.
19. Sol Greear, interview, Clarkesville, GA, July 1995.

NOTES FOR CHAPTER 4: PIONEER TALES

"The Dwelling Place of the Phantom Things That Were..."

1. Much of the information on migration patterns was obtained in a series of interviews with Dale Couch, Georgia Department of Archives and History, Atlanta, 1994-95. He also supplied additional cultural information incorporated into the section on the pioneer settlers of the Helen valley.
2. Based on histories for the England, Pitner, Bell, Conley, Fain, and Vandiver families, and those of other local families contained in *A History of White County* 1857-1980 (Cleveland, GA: White County Historical Society, 1980), "Family Histories of White County", pp. 65-389.
3. Theodore Roosevelt, *Winning Of The West,* vol. 1, chap. V, p. 132, quoted in *The Heritage of Burke County* (Morganton, NC: Burke County Historical Society, 1981), p. 2.
4. Franklin County Historical Society, *History of Franklin County Georgia* (Roswell, GA: WH Wolfe and Associates, 1986), pp. 101, 111-5.
5. Mooney, *Myths Of The Cherokee,* p. 221.
6. E. Merton Coulter, *George Walton Williams: The Life of A Southern Merchant and Banker, 1820-1903* (Athens, GA: Hibriten Press, 1976.), p. 7.
7. Charles Jones, Jr., *History of Georgia, Volume II: Revolutionary Epoch* (Boston: Houghton, Mifflin and Company, 1883; full text reprinted Spartanburg, SC: The Reprint Company, 1965), p. 138.

Pioneer Families of the Helen Valley

1. Much of the narrative is based on extensive review of deeds and census records. Since scores of deed records and numerous census records from 1790-1880 involving Virginia, the Carolinas, Tennessee and Georgia were reviewed, individual notes are employed only in unusual cases. Where possible, these sources are identified in a general way in the text. In addition to property descriptions, deed records give the residence of buyer and seller, and almost always contain other information and clues about previous owners, neighbors, occupations, etc. Since the Helen valley was in Habersham County until 1857 and in White County thereafter, official records from both courthouses were utilized. Copies of 1800's deed records for both counties are on microfilm at the Georgia Department of Archives and History in Atlanta.
2. Max Fain, *Fain Notes,* vol. 1 (Atlanta: n.p., 1981) pp. 83 and 145.
3. One source of extensive information on Jarratt is Robert Bouwman, *Travelers Rest and the Tugaloo Crossroads.*
4. The will is in Habersham County Ordinary Court Minutes Book 4a (1820-50), p. 82. Copy on microfilm at Georgia Department of Archives and History.
5. Unpublished map entitled, "Nacoochee 1837". This map, copies of which are widely held in the Helen-Nacoochee area, originated with R.C. Moffat, who signed it in 1881. It was subsequently embellished (with topographical shading of varying accuracy) by V.R. Hollis, whose name appears on some versions along with the date 1922. Moffat apparently worked in the area as a surveyor before moving or returning to New York state, where he drew the map from memory and any notes he had retained. The map is inaccurate in some small details, but presents a wealth of information which coincides to a high degree with other records from the period.
6. The Old Buncombe County Genealogical Society, *The Heritage of Olde Buncombe County,* vol.1 (Asheville, NC: The Old Buncombe County Genealogical Society, 1981), p. 151.
7. Ibid., p. 149.
8. Genealogical Society of Old Tryon County, *The Heritage of Rutherford County, North Carolina,* vol. 1 (Forest City, NC: Genealogical Society of Old Tryon County, 1984), p. 57.
9. *Heritage of Buncombe County,* p. 149, and 1870 US Census.

10. "Pitner Profiles" Number 2, 1983, unpublished family genealogical bulletin compiled by Donald Pitner Smith. Although there are almost certainly some Georgia descendants, none could be located. Mr. Smith lives in Glen Rock, NJ.

11. Moffat Map.

12. The Moffat Map shows a blacksmith adjacent to the Pitner home. Haynes was listed beside Pitner in the 1840 US census, which showed him engaged in a craft or trade. He was missing from the 1850 census.

13. Criminal Docket "A" containing Habersham Superior Court records for the period 1818-1848 was extracted in Herbert M. Kimzey, *Early Genealogical and Historical Records, Habersham County, Georgia,* compiled by Nancy Kimzey Dempsey (Athens, GA: n.p., 1988), pp. 163-89.

14. George Gordon Ward, *The Annals of Upper Georgia, Centered in Gilmer County* (Ellijay, GA: n.p., 1965), pp. 72-3.

15. Moffat Map.

16. Estate inventories originated in colonial Virginia, and were a standard practice in northeast Georgia until the early 1900's.

17. From England family history and genealogical information provided by England descendant Mrs. Martha Edge of Gainesville, GA.

18. Kimzey, *Early Records,* pp. 170,73,77.

19. Habersham County Ordinary Court Minutes Book 4a (1820-50), p. 134.

20. These trails are shown on a map of the 5th District of Habersham County prepared by the Georgia Surveyor General's office from a survey conducted in 1820 by William Evans. Copies of the map and Evan's notes are on file and microfilm at the Georgia Department of Archives and History in Atlanta.

21. Moffat Map.

22. Burke County Historical Society, *Heritage,* pp. 147-8.

23. A version of this much quoted and often modified list appears in Kimzey, *Early Records,* pp. 5-10.

24. Thomas N. Lumsden, MD, *Nacoochee Valley, Its Times and Places* (n.p., 1989), pp. 25-6.

25. Comments are based in part on information provided by Dale Couch at the Georgia Department of Archives and History and observations relayed by long time area residents Comer Vandiver and Laura Abernathy Cannon.

26. Bouwman, *Traveler's Rest,* pp. 190-6.

NOTES FOR CHAPTER 5: THE DIARIES OF SAM CONLEY

1. Slaves were listed separately in the 1850 and 1860 US Census. The 1860 Slave Census listed the number of slave houses.

2. These diaries were provided by Mrs. Ann Conley of Dallas, Texas, who married Sam's grandson Bill Conley.

3. See, for example, Zell Miller, *The Mountains Within Me* (Toccoa, GA: Commercial Printing Company, 1976), pp. 120-1.

4. Mary Henderson Davidson, a great-granddaughter of Henry Conley, in an interview at her home in Cleveland, GA, December, 1992.

5. George Gordon Ward, *Annals of Upper Georgia,* pp. 72-3.

6. White County 1864 Tax Digest. Copy on microfilm at Department of Archives and History in Atlanta.

7. The last inventory taken in White County was dated June 7, 1926. The final item on the list was "1 gun", valued at $5.00.

8. *Cleveland Courier,* "Honored Memory of Dr. E.F. Starr", 29 June 1923, p. 1.

9. *Clarkesville Advertiser,* advertised schedule of Northeastern Georgia Railroad, 23 October, 1883, pg. 1.

10. *Cleveland Advertiser,* "We Want A Railroad", 24 January, 1880, pg. 1.

11. Lumsden, interview, March 1996.

12. Will of Henry Highland Conley is contained in White County Probate records, Will Book 1, pp. 34-5.

NOTES FOR CHAPTER 6: GOLDEN DREAMS

Mother Nature and the Eastern End of the Georgia Gold Rush

1. See, for example, Larry Otwell, *The Gold of White County Georgia* (Cleveland, GA: Rainbow Sequoia, 1984), pp. 5-6; Andrew W. Cain, *History of Lumpkin County* (Atlanta: Stein Publishing Company, 1932; reprint ed., Spartanburg, SC, The Reprint Company, 1978) pp. 92-4; and David Williams, *The Georgia Gold Rush, Twenty-Niners, Cherokees and Gold Fever* (Columbia, SC: University of South Carolina Press, 1993) pp. 7-13.
2. Williams, *Georgia Gold Rush,* pp. 24-5.
3. Bruce Roberts, *The Carolina Gold Rush: The Nation's First* (Charlotte, NC: Heritage Printers, Inc. ,1982), chap. 1; and William S. Powell, *North Carolina Through Four Centuries* (Chapel Hill, NC: University of North Carolina Press, 1989),pp. 311-2.
4. Coulter, *Auraria,* pp. 4-5. This excerpt taken from James L. Covington, "Letters from the Georgia Gold Region" in *Georgia Historical Quarterly, XXXIX,* 4 (December, 1955): pp. 407-8.
5. Letter from William L. Gwyn to Col. Hamilton Brown, January 31, 1833, in T. Conn Bryan, "Letters Concerning Georgia Gold Mines, 1830-1834", *Georgia Historical Quarterly 44* (1960): p. 343.
6. Letters from William L. Gwyn to Col. Hamilton Brown, in Bryan, "Letters", pp. 342-5. Original letters are preserved in the Hamilton Brown collection, University of North Carolina Library, Chapel Hill.
7. C.H.V.Sutherland, *Gold:Its Beauty, Power, and Allure,* 2d rev. ed. (London: Thames and Hudson, 1969), p. 154.
8. For an overview of the types of gold deposits, see Otwell, *Gold of White County,* pp. 19-26.
9. Jim Vandiver, interview at his Helen home, October, 1995.
10. For example, see Williams, *Georgia Gold Rush,* pp. 65-9.
11. Cain, *History of Lumpkin County,* p. 103.
12. Ibid., p. 94.
13. Stephenson, *Geology and Mineralogy of Georgia,* p. 87.
14. This exact phrase was volunteered by local experts Dr. Tom Lumsden and Jim Vandiver in separate interviews. "The South's Revenge" was achieved in the Helen valley when northern investors lost considerable sums with the "Plattsburg" and "Nacoochee Hills" gold mining companies.
15. Leigh Gedney, interview at his Helen home, November, 1995.

Life and Hard Times at the England Mine

1. The England Mine was originally in Habersham County, but was included in the new county of White when it was formed in 1857. The numerous references to property sales, prices, etc. are from deeds contained in the Habersham County Deed Books through 1857, and from the White County records thereafter.
2. Kimzey, *Early Records,* pp. 172-4.
3. Moffat Map, "Nacoochee, 1837".
4. Langdon Cheves, "Blake of South Carolina" in South Carolina Historical Society, *South Carolina Genealogies: Articles from the South Carolina Historical (and Genealogical) Magazine* (Spartanburg, SC: The Reprint Company, 1983), pp. 83-4.
5. Ibid., pp. 84-5.
6. Kimzey, *Early Records,* p. 176. Prior to Blake's bringing charges, McLaughlin was involved in land dealings where the titles were never transferred to Blake.
7. Cain, *History of Lumpkin County,* pp. 71-2.
8. Coulter, *Auraria,* endnote 41 citing "Acts of Georgia 1834", p. 144.
9. Robert Habersham, power of attorney to Richard Habersham, Habersham County Deed Book N, p. 246.
10. Cain, *History,* p. 72. Professor Cain extracted this quotation from records of the Lumpkin County Inferior Court, 1835.
11. W.S. Yeates, S.W. McCallie, and F.P. King; "A Part Of The Gold Deposits Of Georgia;" aka "The Gold Deposits of Georgia," *Georgia Geological Survey,* Bulletin No. 4. (1896). The sec-

tions of this report on White County were written by F.P. King, and extracted in Otwell, Gold of White County. The reference here is on pp. 170-3 of Otwell.
12. White County Superior Court Records, 1896. Reviewed at White County Courthouse.
13. S.P. Jones, "Gold Deposits of Georgia", *Georgia Geological Survey;* Bulletin No. 19. (1909). The sections of this report on White County mines were extracted in Otwell, *Gold of White County.* This reference is on p. 173 of Otwell.

The Great Mining Ditch of J.R. Dean

1. Austin F. Dean, *A Sketch of the Life of H.H. Dean* (Gainesville, GA: n.p., 1928), pp. 3-4.
2. J.R. identified himself as a resident of Habersham County on a number of deeds written in 1857.
3. King in Otwell, pp. 135-6.
4. Dean, *Life of H.H. Dean,* p. 16.
5. White County Deed Book A, p. 159.
6. King in Otwell, p. 134.
7. King in Otwell, p. 140.
8. Moffat Map, "Nacoochee 1837".
9. Laura Abernathy Cannon, interview at her home in Helen, February, 1992. Laura lived on the site of the Dean home, in a house built after the Dean home burned.
10. Stephenson, *Geology and Mineralogy of Georgia,* pp. 102-3.
11. Ibid.
12. Andrew Cain, *History of Lumpkin County,* p. 110.
13. Ibid., pp. 106-10.
14. King in Otwell, p. 162.
15. Records of White County Superior Court, May term 1880. Reviewed at White County Courthouse.
16. Jones in Otwell, p. 163.
17. Jones in Otwell, p. 142.
18. Dean, *Life of H.H. Dean,* p. 6.
19. Lumsden, *Nacoochee Valley,* p. 23.
20. Laura Abernathy Cannon, interview at her Helen home, February, 1992.
21. Dean, *Life of H.H. Dean,* pp. 15-7.
22. Sybil Wood McRay, *The Tombstone Inscriptions of Hall County, Georgia* (Gainesville, GA: n.p., 1971), p. 15.
23. According to the caretaker of the Alta Vista Cemetery, it is not known exactly when the earliest blocks were opened. Block 2 was apparently opened first, but its number assigned out of order some years later. With the exception of two graves which identify dates of passing decades before Alta Vista was formally established, the earliest burials in Block 3 were in the mid 1880's.

CHAPTER 7: LAURA

Story of the "Ben Fields"

1. Robert Wauchope, *Archaeological Survey of Northern Georgia* (Salt Lake City: The Society for American Archaeology, 1966), pp. 360-3.
2. Ibid., p. 466.
3. Don L. Shadburn, *Cherokee Planters in Georgia, 1832-1838: Historical Essays on Eleven Counties in the Cherokee Nation of Georgia* (Roswell, GA: WH Wolfe Associates, 1990), p. 287.
4. White County Historical Society, *History of White County,* p. 80.
5. King in Otwell, p. 119.
6. King in Otwell, p. 121.
7. King in Otwell, pp. 119-20.

ORDER FORM

Please send _____ copies of LIVING ON THE UNICOI ROAD: Helen's Pioneer Century and Tales From the Georgia Gold Rush. I have enclosed a check or money order for $11.95 per book plus $3.00 per order for shipping and handling.

NAME: _____

ADDRESS: _____

CITY: _____

MAIL TO: Little Star Press
Dept. 212
175 Mt. Calvary Road
Marietta, GA 30064 U.S.A.

Thanks for your order.